AF323159

Lecture Notes on
Computational Structural Biology

Lecture Notes on Computational Structural Biology

Zhijun Wu
Iowa State University, USA

World Scientific

NEW JERSEY · LONDON · SINGAPORE · BEIJING · SHANGHAI · HONG KONG · TAIPEI · CHENNAI

Published by

World Scientific Publishing Co. Pte. Ltd.

5 Toh Tuck Link, Singapore 596224

USA office: 27 Warren Street, Suite 401-402, Hackensack, NJ 07601

UK office: 57 Shelton Street, Covent Garden, London WC2H 9HE

British Library Cataloguing-in-Publication Data
A catalogue record for this book is available from the British Library.

LECTURE NOTES ON COMPUTATIONAL STRUCTURAL BIOLOGY

ISBN-13 978-981-270-589-1
ISBN-10 981-270-589-9

Typeset by Stallion Press
Email: enquiries@stallionpress.com

Printed in Singapore by Mainland Press Pte Ltd

To my family

Preface

This book is a collection of the lecture notes for a course on computational structural biology taught at the Program on Bioinformatics and Computational Biology at Iowa State University, USA. It is, in a certain sense, a summary of what I have learned and understood about computational structural biology based on my own research experience, and is more of a review from the perspective of a computer scientist or applied mathematician on the subject.

The book is focused on introducing computational and mathematical problems, their solution methods, and the research developments. Efforts have been made to keep the contents self-contained and not too technical, so that readers not previously exposed to related subjects can still understand the materials and appreciate the biological importance and mathematical novelty of the research in the field.

The field of computational structural biology is new and still developing. As such, it is hard to define the exact scope of the field, as different groups of researchers may prefer different names with different research goals and scopes in their minds, such as structural bioinformatics, structural genomics, or biomolecular modeling, to name a few. Here, we just choose the current name, not exclusively, and restrict our discussion to the use of computational methods for the study of the structures and dynamics of biomolecules and, in particular, proteins.

Structures and their dynamic properties are key to the understanding of the biological functions of proteins. However, in general, they are difficult to determine. Both experimental and theoretical approaches have been used. They all require intensive mathematical computing, including data analysis, model building, structure prediction, and dynamics simulation. Mathematical problems arise in various stages of investigations on structure and dynamics. Some of them can be resolved effectively by using existing mathematical tools, while others present remarkable computational challenges. The development of computational structural biology thus follows the great need for mathematics and computation in structural research.

In Chap. 1, we discuss the importance of structures for the understanding of proteins and related biological functions. Two classes of computational problems, structure determination and dynamics simulation, as the main subjects of computational structural biology, are introduced. A review on the famous problem of protein folding is given.

The topics related to X-ray crystallography and nuclear magnetic resonance (NMR) spectroscopy, two major experimental approaches for structure determination, are covered in Chaps. 2 and 3, respectively, with emphasis on the phase problem in X-ray crystallography and the distance geometry problem in NMR and their solutions. Two important computational methods, potential energy minimization and molecular dynamics simulation, as well as their extensive applications to protein structures and dynamics are discussed in Chaps. 4 and 5.

The knowledge-based approach to structure prediction, an active area of research in recent years, is reviewed in Chap. 6, with a great deal of details left for readers to explore further. The development of other related fields, including computational genomics and systems biology, and their connection with computational structural biology are also discussed.

To be self-contained, some general knowledge on algorithms and numerical methods are given with relatively rigorous mathematical descriptions in Appendices A and B. In contrast, in the main chapters of the book, the presentation is made purposely without formal mathematical statements (such as theorems and proofs). The bibliography is provided in the form of further readings attached to each chapter and grouped under specific topics. They are selected readings, not complete references, and are also given in certain suggested reading orders.

I would like to thank my students and colleagues, Ajith Gunaratne, Feng Cui, Qunfeng Dong, Rahul Ravindrudu, Vladimir Sukhoy, Peter Vedell, Atilla Sit, Xiaoyong Sun, Di Wu, and Wonbin Young, for sharing their knowledge and providing valuable insights while working with me. The constant discussions among us on research in computational biology have been the major source of motivation for the writing of the book. I would also like to thank my biology colleagues David Shalloway, George Phillips, and Robert Jernigan for working with me and introducing me to the fascinating field of computational biology; and my optimization advisers and colleagues John Dennis, Robert Bixby, Tom Coleman, Jorge Moré, Richard Tapia, Yin Zhang, Guangye Li, Yinyu Ye, and Yaxiang Yuan for teaching me optimization and encouraging me to explore scientific applications. I would like to acknowledge the strong support provided by the Department of Mathematics, the Graduate Program on Bioinformatics and Computational Biology, and the Baker Center for Bioinformatics and Biological Statistics, Iowa State University. I have also benefited from the support provided by the Mathematical Biosciences Institute at Ohio State University, USA, the Institute for Mathematics and Its Applications at the University of Minnesota, USA, and the Institute of Computational Mathematics at the Chinese Academy of Sciences, China, for allowing me to visit the institutes while finishing the book. Finally, I would like to thank Peng Wu for reading the initial draft of the book and providing helpful suggestions; my friend and colleague Dr Chunguang Sun at World Scientific Publishing Company (WSPC) for suggesting the book project; and the WSPC staff, Wanda Tan, Azouri Bongso, and Mei-Lian Ho, for their professional help and friendly cooperation on text organization, cover design, and proofreading.

Z. Wu

Contents

Chapter 1

Introduction

Proteins are an important class of biomolecules. They are encoded in genes and expressed in cells via genetic translation. Proteins are life-supporting (or sometimes, destructive) ingredients and are indispensable for almost all biological processes. In order to understand the diverse biological functions of proteins, knowledge of the three-dimensional (3D) structures of proteins and their dynamic behaviors is essential. Unfortunately, these properties are difficult to be determined either experimentally or theoretically. The goal of computational structural biology is to provide an alternative, or sometimes, complementary, approach to protein structures and dynamics by using computer modeling and simulation.

1.1. Protein Structure

A protein consists of a sequence of amino acids, typically several hundreds in length. There are 20 different amino acids. Therefore, millions of different proteins can be formed with different amino acid sequences and often different functions. In the human body alone, there are at least several hundreds of thousands of different proteins, with functions ranging from transporting chemicals to passing electrical signals, from activating cellular processes to preventing foreign intrusion, and from forming all kinds of molecular complexes to supporting various physical structures of life.

1.1.1. *DNA, RNA, and protein*

Not all sequences of amino acids are biologically meaningful. Those used in proteins are selected or decided by the biological systems. They are encoded as genes in the DNA sequences and expressed in the cells at certain times and places. A typical gene expression process occurs as follows: a gene, as a DNA sequence, that encodes a protein is first transcribed into a corresponding RNA sequence; the RNA sequence is then translated into an amino acid sequence required by the protein (Fig. 1.1).

A DNA sequence is made of two complementary chains of four different deoxyribonucleic acids — known as adenine (A), cytosine (C), guanine (G), and thymine (T) — with A in one of the chains pairing with T in another, and C pairing with G. The two chains wrap around each other and form a so-called double helix. During the transcription process, the double helix unwinds, and one of the strands is used as a template to make a single chain of RNA, which consists of four ribonucleic acids — A, C, G, and U (uracil) — that correspond to the deoxyribonucleic acids — A, C, G, and T, respectively — in the DNA templates. In this sense, the RNA sequence is equivalent to the DNA sequence; it is only transcribed in a different form. The RNA

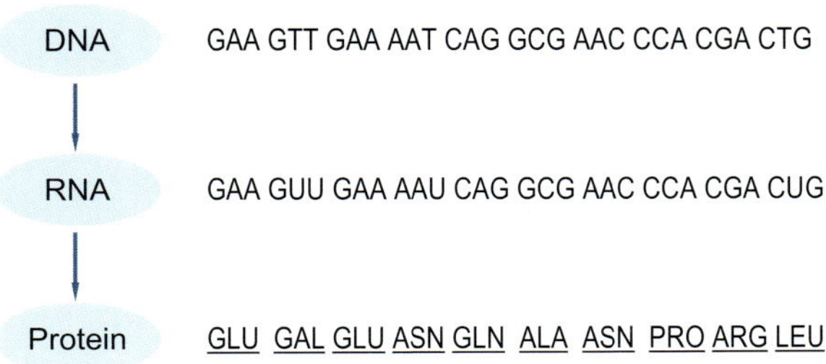

Fig. 1.1. Central dogma of molecular biology. A DNA sequence is transcribed into an RNA sequence, and an RNA sequence is translated into an amino acid sequence, which forms a polypeptide chain and folds into a protein.

(Second Base)

	U	C	A	G	
U	Phe Phe Leu Leu	Ser Ser Ser Ser	Tyr Tyr End End	Cys Cys End Trp	U C A G
C	Leu Leu Leu Leu	Pro Pro Pro Pro	His His Gln Gln	Arg Arg Arg Arg	U C A G
A	Ile Ile Ile Met	Thr Thr Thr Thr	Asn Asn Lys Lys	Ser Ser Arg Arg	U C A G
G	Val Val Val Val	Ala Ala Ala Ala	Asp Asp Glu Glu	Gly Gly Gly Gly	U C A G

(First Base) — (Third Base)

Fig. 1.2. Genetic code. A sequence of three RNA molecules translates into one amino acid; there are 20 different amino acids that can be translated from RNA.

chain is processed to produce a chain of amino acids, with every three contiguous ribonucleic acids used to make one amino acid following a generic translation code (Fig. 1.2). There are a total of 64 different RNA triplets. They are mapped to 20 different amino acids. Two amino acids make a dipeptide. The chain of amino acids generated from the RNA chain makes a polypeptide. It folds into a unique 3D structure to become a functional protein.

Here, the DNA sequence, or the gene, determines the RNA sequence, and the RNA sequence then determines the amino acid sequence. However, the amino acid sequence, or the polypeptide, will not function as a normal protein until it folds into an appropriate 3D structure called the native structure of the protein. The latter step is called protein folding, and has been a fundamental research topic in biology because of its obvious importance in the study of proteins.

1.1.2. *Hierarchy of structures*

When forming the 3D structure, the chain of amino acids of the protein is not broken. Neighboring amino acids in the chain are always connected by strong chemical bonds. This connected chain of amino acids forms the primary structure of the protein. Different parts of the chain may form different types of structures, depending on their amino acid sequences and sometimes their interactions with other parts of the chain. The most commonly seen structures are α-helices and β-sheets. An α-helix is a helical type of structure, most often right-handed. Usually, about 3.6 amino acids form one circular section of the helix, with a 5.4 Å elevation on average. A β-sheet is a pleated sheet type of structure formed by a group of chain sections stretched and aligned in parallel. These structures are called the secondary structures of the protein. The secondary structures assemble themselves to finally form the overall structure called the tertiary structure of the protein (Fig. 1.3).

The primary structure shows the connectivity of the amino acids in sequence. It contains all of the information that defines the protein and hence its structure. At this level, the protein appears as a linear

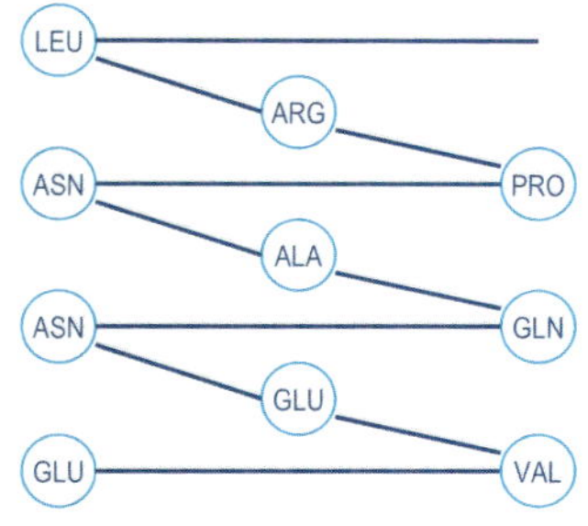

Fig. 1.3. Protein folding. A polypeptide chain becomes a functional protein after it folds into a proper 3D structure; the latter is called the native structure of the protein.

chain of amino acids. The secondary structure demonstrates the protein's local structural patterns. The protein can then be viewed at a higher level as a set of α-helices and β-sheets pieced together. The segments between secondary structures usually do not have regular forms; they are called loops or random coils. The tertiary structure gives the detailed arrangement of all the segments of the protein in 3D space. It is usually a compact structure, and is unique for each protein. Most importantly, it is physically or biologically the most favorable structure that the protein takes and is a structure necessary for the protein to achieve its physiological function.

1.1.3. *Properties of amino acids*

Amino acids are the building blocks of proteins. The properties of the amino acids enable proteins to form certain types of secondary and tertiary structures and to have various biological and chemical functions.

There are 20 different amino acids. They all have an amino group (H_2N), a carboxyl group (COOH), and a side chain (R). A carbon atom called C_α connects these three parts. The side chains are different for different amino acids. Except for glycine whose side chain has only a hydrogen atom, all side chains have a carbon atom called C_β connected to C_α.

When two amino acids are connected to form a dipeptide, the carboxyl group of one amino acid interacts with the amino group of another and makes a covalent bond called the peptide bond. Each amino acid part between two peptide bonds in a polypeptide chain is also called a residue. Since the connection happens only between the amino group and the carboxyl group of the neighboring amino acids, the amino acid chain without the side chains is called the main chain or the backbone of the protein.

Within each amino acid residue, some local structures are relatively easy to be determined and they do not change. For example, the bond lengths and bond angles can be determined based on the chemical knowledge of the bonds and are relatively fixed. Therefore, the amino group and the carboxyl group, as well as some parts of the side chains, can be determined and considered as rigid. However, the bonds connecting C_α to the functional groups and some of the bonds in the side chains are flexible and can rotate. The angles around the flexible bonds are called the dihedral angles or the torsion angles. They are the main structural freedoms to be determined for the formation of correct protein folds.

Amino acids are not just the building material of proteins, but are also the driving force for proteins to form different levels of structures. For example, the hydrogen bonds between amino acids in the neighboring coils of α-helices or in the neighboring strips of β-sheets are the key factors for proteins to form these secondary structures. Also, the fact that amino acids can be either hydrophobic or hydrophilic and proteins tend to form tertiary structures with hydrophobic cores has been considered as one of the guiding principles of protein folding.

Amino acids can be grouped according to the hydrophobic or hydrophilic properties of their side chains. One group includes the amino acids with nonpolar side chains, which make the amino acids hydrophobic. The hydrophobic group contains glycine, alanine, valine, leucine, isoleucine, methionine, phenylalanine, tryptophan, and proline. Another group includes the amino acids with polar side chains, which make the amino acids hydrophilic. The hydrophilic group includes serine, threonine, cysteine, tyrosine, asparagine, and glutamine. Acidic amino acids are those with side chains that are generally negative in charge because of the presence of a carboxyl group, which is usually dissociated (ionized) in cellular pH. They include aspartic acid and glutamic acid. Basic amino acids have amino groups in their side chains that are generally positive in charge. They include lysine, arginine, and histidine. These last two types of amino acids may be considered as electrically charged amino acids, and are hydrophilic.

1.1.4. *Sequence, structure, and function*

There are close connections among sequences, structures, and functions. Sequences determine structures, and structures determine functions. In a broader sense, this implies that similar sequences often have similar structures, and that similar structures often function similarly, although conversely it is not always so. From the point of view of evolution, genes mutate as biological systems develop. Genes that belong to close families are certainly similar and the proteins expressed from them should carry similar structures as well. The conserved parts of the structures are very likely to correspond to some biological functions shared by the similar genes and hence the proteins. These relationships among sequences, structures, and functions are fundamental questions investigated in modern molecular genetics. They are important properties that are often employed in structural and functional studies.

Here, we look at two example proteins — a retrotranscriptase found in the HIV virus (Fig. 1.4) and a prion protein that causes mad cow disease (Fig. 1.5) — and see what their sequences, structures, and functions are and how they are related.

The HIV virus is a retrovirus, meaning that it is composed of an RNA sequence. Once the virus invades a cell, the RNA sequence is transcribed back to a DNA sequence, which then integrates with the

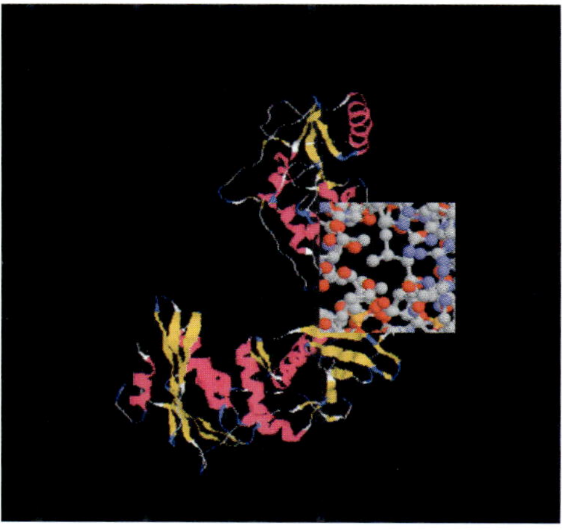

Fig. 1.4. HIV retrotranscriptase. The enzyme transcribes virus RNA back to DNA so that the virus can be integrated in the host genome; it is 554 residues long and has 4200 atoms without counting the hydrogen atoms.

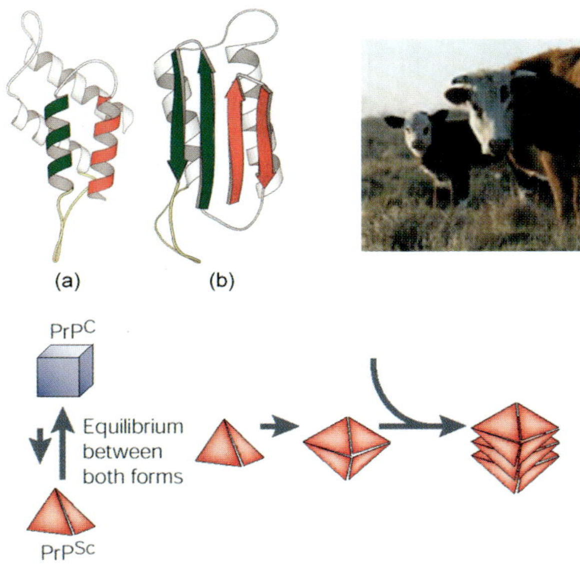

Fig. 1.5. Prion transformation. A normal prion (PrP^C) can be transformed into an abnormal form (PrP^{Sc}). The latter may be aggregated to dysfunctional complexes that can damage neuron cells and cause mad cow disease.

cell's normal genome so that the viral RNA can be reproduced. Here, the RNA-to-DNA transcription is called retrotranscription, and it occurs only in the presence of a protein called the retrotranscriptase. It turns out that the viral RNA itself contains a gene that produces the required retrotranscriptase. Retrotranscriptase is generated when this gene is expressed in the cell. The former binds onto the viral RNA and transcribes it into a DNA sequence and so on and so forth. The transcriptase protein has a special 3D structure so that it can bind properly onto the virus. The function of the protein therefore depends heavily on its 3D structure. On the other hand, knowing this dependency, researchers have been able to develop drugs that can destroy the 3D structure of the protein and obstruct its proper function, thereby preventing the virus from being transcribed. This is an example of how the structure of a protein is related to its function, and how it can sometimes be used for medical purposes.

The prion protein has been studied extensively, not only because it is related to the cause of mad cow disease, but also because it gives rise to a rare case of protein infection rather than the conventional viral or bacterial infection. The prion gene is short and expresses a small protein with around 200 residues. The normal prion protein exists in regular cells. Its function is not well understood. It folds into a simple structure with two α-helices and three β-sheets. However, the protein sometimes misfolds into a structure with many β-sheets. The latter may affect a normal prion, causing it to also take on the same misfolded shape. If the process continues, a large number of misfolded prions accumulate and can damage the cells, in particular the neurons in the brain, causing mad cow disease. This is an example of how structure can affect a protein's function when it is not folded properly, and again how important this can be in medical research.

1.2. Structure Determination

With the completion of the genomic sequencing of human and many other species, studies on proteins, the end products of gene expression, have become urgently more important for the interpretation of genes and their implications on life. However, to understand proteins and their functions, it is essential to know their 3D structures, which, due to various technical reasons, are generally difficult to be determined. This has therefore created a research bottleneck yet to be overcome.

1.2.1. *Experimental approaches*

There is no direct physical means to observe the structure of a protein at a desired resolution, for example, at the residue level. Several experimental approaches have been used to obtain some indirect structural data upon which the structures may be deduced. For example, the diffraction data for a protein crystal can be obtained by X-ray crystallography, and can be used to find the electron density distribution and hence the structure of the protein; in addition, the magnetic resonance spectra of the nuclear spins in a protein can be detected by nuclear magnetic resonance (NMR) experiments and can be used to estimate the distances between certain pairs of atoms, and subsequently the coordinates of the atoms in the protein. In both cases, computation plays an important role in collecting and analyzing data and in forming the final structures.

X-ray crystallography and NMR spectroscopy are major experimental techniques used for structure determination. Surveys on the protein structures deposited in the Protein Data Bank (PDB) show that 80% of the structures were determined by X-ray crystallography, 15% by NMR, and 5% by other approaches. These structures, about 30 000 in total, contain a high percentage of replications (structures for the same protein determined with different techniques or under different conditions). Some structures are also very similar because there are only few mutations among them. Without counting the replications and the genetically highly related structures, there may be only around several thousands of different proteins whose structures have been determined. However, there are at least several hundreds of thousands of different proteins in the human body alone. Most of their structures are still unknown.

The experimental approaches have various limitations. For example, X-ray crystallography requires crystallizing the protein, which is time-consuming and often fails. To obtain accurate enough signals, NMR experiments can only be carried out for small proteins with less than a few hundred residues. Therefore, the number of structures that can be determined by these experimental approaches is far from adequate with respect to the increasing demands for structural information on the hundreds of thousands of proteins of biological and medical importance.

1.2.2. *Theoretical approaches*

Theoretical approaches have been actively pursued as complementary or alternative solutions for structure determination. They are based on physical principles and therefore also called *ab initio* approaches. For example, with molecular dynamics simulation, one can compute protein motion by solving a system of equations of motion in the force field of the protein. Then, ideally, if the entire motion of protein folding can be simulated, the structure that the protein eventually folds into may be obtained when the simulation reaches the equilibrium state of the protein. Unfortunately, folding is a relatively long process. The simulation requires a sequence of calculations which is so lengthy that the simulation cannot be completed in a reasonable amount of time, using current simulation algorithms and available computers.

An alternative approach is to disregard the physical process of folding and directly search for the protein's native fold in the protein's conformational space by using an optimization algorithm. This approach assumes that the structure into which the protein folds has the least potential energy. Therefore, a structure that minimizes the potential energy of the protein is sought. The native fold of the protein should correspond to the global minimum of the potential energy. Here, similar to molecular dynamics simulation, where the force field must be given for the protein, a potential energy function needs to be defined so that the potential energy of the protein can be calculated (and optimized) accurately. However, even if such a function was available, it could still be difficult to find the global minimum of the function.

Work has been done and progress has been made in the development of efficient algorithms so that the simulation of protein folding and the search for the global minimum of protein potential energy may eventually become feasible. For example, algorithms have been developed to increase the step size so that large-scale motions can be simulated in a given time. Furthermore, information on the end structure has been used to guide the simulation. Parallel computation has been employed to accelerate the simulation. Additionally, reduced protein models have been introduced so that dynamics simulation and global optimization may be performed more effectively and efficiently at a level higher than the atomic level.

1.2.3. *Knowledge-based methods*

There is another group of active approaches for structure determination that can be classified as theoretical, but is instead grouped separately as knowledge-based because they are based on the knowledge of known protein structures rather than physical theories. In these approaches, proteins are analyzed and compared with proteins of known structures at either the sequential or structural level. Their structures can then be modeled by using the known protein structures as templates. Because the proteins need to be compared with the existing ones, these approaches have also been called comparative approaches.

The theory behind the comparative approaches is that genetically closely related proteins should have similar functions and structures. Therefore, if two proteins have similar sequences of amino acids, their structures should look alike, and one may be used as a template for another. The approaches that rely completely on genetic similarities among proteins are also called homology modeling, because the genetically closely related proteins are called homologs. Sequence comparison is not hard in general, although the sequences need to be properly aligned and appropriate scoring functions have to be developed for the evaluation of the similarity between two given sequences.

Not all similar proteins correspond to similar sequences. In fact, there are many cases where different sequences lead to similar proteins structurally and functionally. In order to find such similarities among proteins and utilize them for structural modeling, more information is required rather than just sequences. Various techniques including structural alignment have been developed to compare proteins in terms of both sequential and structural similarities. They have been applied to inverse folding as well, i.e. to find all sequences or proteins that can fold into a given structure (as opposed to finding the sequences or proteins of known structures with which a given sequence or protein may share a similar fold).

1.2.4. *Structural refinement*

Due to experimental errors, the structures determined by X-ray crystallography or NMR spectroscopy are often not as accurate as desired. Further refinement of the structures, including human intervention, is always required. For example, based on the electron density distribution generated by X-ray crystallography, the positions for the atoms in the protein can be assigned, but they are usually not very accurate. One way to improve the structure is to use an empirical energy function for the protein to adjust the positions of the atoms so that the energy of the protein can be minimized.

In NMR, in addition to obtaining additional experimental data to further improve the structure, several iterations may be required to obtain an ensemble of structures that can eventually satisfy the experimental distance data. The latter may require the solution of a nontrivial mathematical problem called the distance geometry problem. Nonetheless, the experimental data may not be sufficient for the complete determination of the structure. A subsequent energy minimization step may also be necessary. Even so, many NMR structures are still not as accurate and detailed as X-ray structures. Further justification of the structures remains an important research issue.

The refinement of comparative models presents even greater challenges. First, the structures can provide correct models for most but not all local regions of proteins, even if they are obtained from proteins of high sequence and structural similarities. Second, without further experimental evidence, it is difficult to judge whether a model is truly correct or there is still much to improve. If a model is indeed close to the true structure, refinement will be possible; otherwise, further improvement of the model will not differ much from *ab initio* structural determination. A refinement algorithm that can only provide small improvements on the model will not be of much help.

1.3. Dynamics Simulation

A protein structure changes dynamically, not only during folding, but also at equilibrium. The conformational changes of a protein over a certain time period are called the general dynamic properties, while the average behaviors of the protein around an equilibrium state are called the thermodynamic properties. In many cases, the dynamic and thermodynamic properties of proteins are just as important as their structures for the study of proteins, but they are even harder to examine experimentally.

There are two ways of conducting protein dynamics simulation. One is of a more stochastic nature, using the so-called Monte Carlo method to obtain a large number of random samples of physical states. The general behaviors of the protein are then evaluated with the samples. Another is based on the simulation of the motions of the particles, namely, atoms, in the protein through the solution of a time-dependent system of equations. A phase space trajectory can be obtained from the solution, with which various dynamic properties of the protein can be analyzed.

1.3.1. *Potential energy and force field*

A force field can be defined based on the physical interactions among the atoms in a protein. For a given conformation, the force on each atom can then be computed as a function of the positions of all the atoms. A potential energy function can also be defined such that the negative derivatives of the function with respect to the positions of the atoms equal the corresponding forces on the atoms. The potential energy and force field must be computable one way or another in order to perform dynamics simulation.

If the forces for all the atoms are equal to zero, the atoms will stop moving. The protein is said to be in an equilibrium state. In an equilibrium state, the protein either stays in one conformation or, in most cases when the kinetic energy is not equal to zero, oscillates around the equilibrium state. Since the forces are the negative derivatives of the potential energy, at the equilibrium state, the potential energy is at a stationary point, most likely an energy minimum. At an energy minimum, the protein should be relatively stable because it has the lowest energy at least within a small neighborhood around the minimum. In general, it is assumed that the protein native conformation has the lowest potential energy in the entire conformational space or in other words is a global energy minimum, and therefore should be in the most stable state.

In principle, the potential energy for a molecule can be computed with quantum chemistry theory. However, the required calculation increases rapidly with an increasing number of atoms and electrons in the molecule. A protein may have several thousands of atoms and several magnitudes more of electrons. Therefore, it is not possible to use the quantum chemistry principle to obtain the potential energy for the protein. A general approach is to use semiempirical functions to calculate the potentials for the protein approximately. The accuracy or the quality of the functions depends on the parameters chosen for the functions. They require carefully collected experimental data to approximate.

In fact, whether or not a protein is in the most stable state should be evaluated in terms of free energy rather than simply the potential energy, because free energy describes more accurately the thermostability of a system. The free energy of a system is in general the internal energy of the system minus the product of the temperature and the entropy. So, at a fixed temperature, the free energy may not be minimized at the potential energy minimum since the entropy may increase the free energy. However, since the entropy is difficult to be evaluated, in practice, only the potential energy is considered as an approximate assessment of the stability of the protein.

1.3.2. *Monte Carlo simulation*

A Monte Carlo approach to dynamics simulation assumes that the physical state of the system to be simulated is subject to the Gibbs–Boltzmann distribution. In other words, if $E(x)$ is the potential energy of the system and x is the state variable, then the probability of the system to stay at state x is $p(x) = \exp[-E(x)/k_B T]/Z$, where k_B is the Boltzmann constant, T is the temperature, and Z is the normalization constant for the distribution.

Based on this assumption, the dynamic behaviors of a physical system can be simulated by generating a large set of sample states that are consistent with the Gibbs–Boltzmann distribution of the system. The statistical properties of the system can then be easily obtained from the samples. Note that the sampled states reflect the probability distribution of the system, but not the dynamic changes over time; hence, they are time-independent. The statistical significance certainly depends on the sufficiency of the sampling. For a large system such as a protein, the sufficiency is sometimes hard to be achieved.

The Monte Carlo simulation is temperature-specific. At a given temperature T, a random state x is generated. The state x is evaluated and taken with a probability $p(x)$ as defined earlier. After the state is evaluated, the next state is generated and tested again, and the whole process is repeated. In this way, the set of accepted states will be subject to the Gibbs–Boltzmann distribution of the system. Because the normalization constant does not change for a given system, the probability function can be calculated without dividing Z. At each step, the next state is usually generated by making a small perturbation on the current state, simulating the change of the system from one state to another.

1.3.3. *Solution of equations of motion*

The major simulation scheme to be discussed in this book is based on the solution of the equations of motion that can be established for the particles in a physical system or, more specifically, the atoms in a molecule. The principle comes from classical mechanics as it is applied to the atoms in the molecule. For each atom i, assuming that Newton's law of motion holds, the mass m_i times the acceleration a_i of the atom should be equal to the force f_i, i.e. $m_i a_i = f_i$, $i = 1, 2, \ldots, n$. Given the fact that f_i is a function of the state variables $\{x_i\}$, with x_i being the position vector of atom i and a_i the second derivative of x_i with respect to time, $m_i d^2 x_i / dt^2 = f_i (x_1, x_2, \ldots, x_n)$.

The above system may have an infinite number of solutions. Only if some additional conditions are imposed, will it have a unique solution. The system may have thousands of equations for proteins. The right-hand side functions are also highly nonlinear. Therefore, solving the system of equations is not trivial. The only way to approach it is to solve it numerically. But still, the simulation can be very time-consuming or in other words computationally very demanding, because millions or even billions of steps are required to complete the simulation of some motions of biological interest, where each step involves the calculation of the positions of all the atoms at a particular time. The reason that the simulation requires so many steps is that the time step has to be very small on the order of femtoseconds to guarantee the accuracy or convergence of the calculations, while many protein motions are on the order of milliseconds or seconds. Because of this nature, protein dynamics simulation remains a challenging research subject to ultimately become a computational tool accessible to most dynamic behaviors of proteins of biological importance.

There are two types of conditions under which dynamics simulation can be performed: initial conditions and boundary conditions. Initial conditions are usually the initial positions and velocities of the atoms, while the boundary conditions are the initial and ending positions of the atoms. The first set of conditions is used to find protein motions with a given initial conformation and temperature. A solution trajectory can be obtained to indicate how the protein conformation changes along a certain pathway under the given initial condition. The second set of conditions is used to find a possible trajectory for the protein motion between two given conformations, for example, how a protein transits from one state to another. Both sets of conditions have some important biological applications. However, the system may be easier to be solved for the first set of conditions. The system under the second set of conditions may have either multiple solutions or no solution at all. The solution method is more complex in general, but is easier to implement on parallel computers, which can be an advantage for large-scale applications.

1.3.4. *Normal mode analysis*

Besides the regular trajectory calculation, an important topic in dynamics simulation is the evaluation of the structural fluctuation of a protein around its equilibrium state, which is often called normal mode analysis. A protein further changes its conformation even after reaching its equilibrium state. The reason is that although the potential energy is minimized, the protein still has kinetic energy, which drives the system away from equilibrium, just like a simple harmonic oscillator vibrating around its equilibrium position.

The fluctuation is different along different directions in conformational space. A linearly independent set of directions can be found for the protein via normal mode analysis, each with a different vibration frequency called the normal mode. Importantly, in contrast to regular trajectory calculation, normal mode analysis can be performed analytically based on the singular value decomposition of the Hessian matrix of the potential energy function at the equilibrium state. The largest singular value corresponds to the fastest vibration frequency.

The vibration of each individual atom is a linear combination of the vibrations in all of the modes. However, if only a few slow modes are included, the fluctuations at a coarse level can then be observed without detailed fast vibrations. Such dynamic properties are often useful for the study of the most important motions of a system. In fact, the slow motions are not affected so much by the fast modes. Therefore, by using only a few slow modes, the size of the variable space for the protein is significantly reduced.

The structural fluctuations of a protein around its equilibrium state can also be obtained by averaging the structural changes obtained in regular dynamics simulation. They can also be derived from Monte Carlo simulation by averaging the fluctuations in the entire ensemble of sampled states. Additionally, there are coarse-level approximation methods such as the Gaussian network model to evaluate fluctuations at only the residue level, with an approximate elastic network model for the residue–residue interactions.

1.4. The Myth of Protein Folding

Having puzzled scientists for decades, the problem of protein folding remains a grand challenge for modern science. While it is a fundamental problem in biology, its solution requires knowledge beyond the traditional field of biology and has motivated research activities across many other disciplines including mathematics, computer science, physics, and chemistry.

1.4.1. *Folding of a closed chain*

First, let us consider the simple mathematical problem of folding an open or closed chain. Suppose that we have a chain with different lengths of links. We call the chain open if the two ends of the chain are not connected; otherwise, we call it closed. It is easy to fold a chain into a line (like folding a necklace into a thin box) if it is open. However, it is not as trivial to fold if the chain is closed. For example, it is harder to put a necklace in a thin box if it is still locked.

Mathematically, folding an open chain is equivalent to finding the positions for a sequence of points (connections in the chain) in a real line so that the distances between the neighboring points are equal to the given lengths (links in the chain). The problem of folding a closed chain has only one more condition that also requires the distance between the first and last points to equal zero. It turns out that this last condition makes the problem harder to solve (Fig. 1.6).

In terms of computational complexity, the problem of folding an open chain can be solved in polynomial time, while the problem of folding a closed chain has been proven to be NP-complete. In other words, an open chain can be folded efficiently in order of n^r steps,

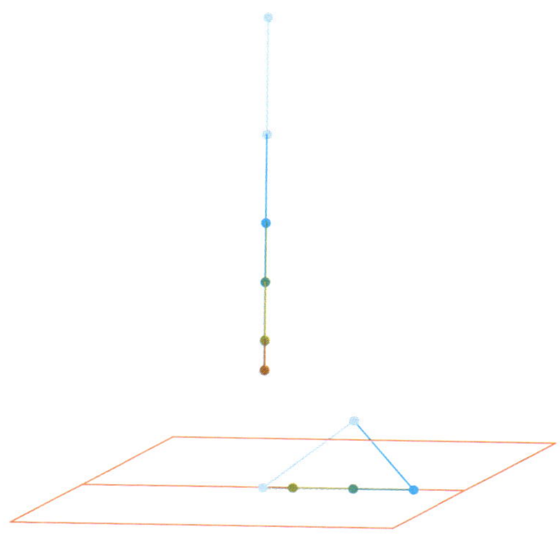

Fig. 1.6. Chain folding. The problem of folding a closed chain is NP-complete.

where n is the number of links in the chain and r is a finite number (e.g. $r = 1$). On the other hand, this can be done for a closed chain only if it is on a nondeterministic computer (nondeterministic polynomial). It would run into exponentially many steps if it is on a regular (deterministic) computer, like what we have.

Folding a protein is certainly a much more complicated problem, but it indeed seems like folding a chain, only with more complex conditions. It is difficult or perhaps irrelevant to show whether protein folding is NP-complete. However, since even folding a closed chain is so hard in general, the problem of protein folding cannot be trivial. Indeed, the solution of the problem has presented great computational difficulties yet to be overcome.

1.4.2. *Biological and physical basis*

Although there may be many factors, including small molecules like chaperones, that affect protein folding, it is agreed that the sequence of amino acids which form the protein determines solely the structure it folds into. In other words, the information that is required for a protein to fold into its structure is fully contained in its constituent amino acids and their order in the sequence. This fact can also be understood to mean that under normal biological or physical conditions, the same sequence of amino acids will fold into the same protein, while other factors may affect only the rate of folding but not the ultimate structure.

The folding of a protein into a 3D structure is a result of the physical interactions among the amino acids in the protein sequence or, more accurately, among the atoms in the protein. The interactions are driven by the potential energy of the protein. The atoms interact to seek a way of lowering the total potential energy. It is assumed that a protein reaches the global potential energy minimum when forming the native structure. The structure is therefore also considered the most stable. Of course, at nonzero temperatures, the free energy instead of potential energy should be used in the assumption, because in general it is the free energy that every physical system tends to minimize.

Depending on the potential energy landscape, folding can be a complicated process. There is a theory that folding takes a unique pathway for each protein, crossing a series of intermediate states. The most recent theoretical and experimental studies have shown that the protein potential energy landscape resembles a funnel shape, with many local minima along the walls of the funnel but the global minimum at the bottom. The local minima along the walls should be easy to skip, while there are still many local minima at the bottom around the global minimum, making the global minimum hard to be found. Also, because of such an energy landscape, there is so far no strong evidence supporting unique folding pathways. Therefore, a common consensus is that folding may have multiple pathways, as it seems possible to start from any point of the funnel and follow a path to reach the bottom of the energy landscape.

1.4.3. *Computer simulation*

The process of folding is hard to be observed experimentally. The only properties that can be observed in physical experiments may be the initial and ending structures and the time for folding. Computer simulation is the only way to determine the folding pathway based on physical principles. To make the computation affordable, force field functions have been developed so that the forces for various atomic-level interactions in proteins can be calculated relatively easily and precisely. By using these force functions, we can in principle simulate protein motions such as protein folding by solving a system of equations of motion for any protein of interest.

There are many methods to solve the equations. The basic idea is to start with the initial positions and velocities of the atoms and then try to find the positions and velocities of the atoms at later times. At each time point, the positions and velocities of the atoms are calculated based on their previous positions and velocities, and the time is advanced by a small step. However, the main challenge is that the step size has to be small — in the order of femtoseconds — to achieve the desired accuracy, while protein folding requires milliseconds or even seconds to complete. Clearly, the simulation of folding may take millions of billions of steps, which can hardly be affordable on even the most powerful computer to date.

Duan and Kollman (1998) performed a 1-μs simulation for a small protein, the chicken villin headpiece subdomain HP36. This protein has 36 residues, with 3000 water molecules added into the system. The simulation was carried out on a 256-processor CRAY T3E, using the software AMBER for the calculation of the force field. The CRAY T3E is a very powerful supercomputer. Even with one processor, it can provide much more computing power than a regular workstation. Still, the entire simulation took 100 days to be completed.

The 1-μs computer simulation of Duan and Kollman (1998) was able to reveal the major folding pathway of HP36, which usually folds in about 10–100 μs. The final structure that the simulation converged to showed a close correlation with the structure obtained through NMR experiments. More importantly, based on the simulation, a folding trajectory was obtained and many intermediate states were discovered. The work was indeed exciting because it was the first protein structure folded on a computer using solely physical principles. It was also the first time that a complete folding pathway was plotted and analyzed.

Efforts have also been made to utilize loosely connected networks of computers for protein folding simulation. Most notable is an Internet website, folding@home, developed to perform folding simulation over the Internet. The idea is that over the Internet, there are hundreds of thousands of computers often in an idle state; therefore, folding@home organizes available computers to donate time for some simulation tasks. For this purpose, the simulation is conducted as follows.

A set of trajectories is first followed by a group of available computers on the Internet. Once in a while, if a faster folding trajectory is found, a new set of starting points is created around the end point of the trajectory, and the computers are stopped to follow the trajectories started with the new points. The process continues until a folding pathway is found by connecting a sequence of restarted trajectories. During the simulation, a computer can participate at any time and request an unfinished task. It can return the task to the system whenever the time is up. The returned task may be continued again when another computer becomes available.

The website folding@home has run for several years, during which quite a few proteins have been folded on the Internet. For example, the folding of the protein HP36 was simulated by Duan and Kollman (1998) for only 1 μs, but the real folding was estimated to take 10–100 μs. It turns out that folding@home was able to complete the entire simulation and provide a full pathway description for folding the protein. The results from folding@home may not yet deliver completely satisfactory answers to the questions of protein folding, but they have shown some promising directions that may lead to an ultimate computational solution to the problem of protein folding if proper models, algorithms, and computing resources can be developed and utilized.

1.4.4. *Alternative approaches*

A computational bottleneck in fold simulation is that a long sequence of iterative steps must be carried out, where each step can start only after the previous one has completely finished. In each step, there are not many calculations to be performed; and even with the most powerful computer (such as a parallel computer), the speedup will be limited and the simulation cannot be completed in a reasonable time. Two alternative approaches to the folding problem are worth mentioning: the boundary value formulation and potential energy minimization approaches.

The first approach, which assumes the availability of the end structure, finds a trajectory that connects the given initial and ending structures. The approach cannot be used if the ending structure is unknown. In such a case where the ending structure is available, the trajectory can be formulated as a boundary value problem, which can be done differently from an initial value problem and in a more parallel fashion than the conventional sequential manner. The latter property makes it possible to perform a complete fold simulation if a massively parallel computer is used.

The second approach sacrifices the attempt to obtain the folding pathway and focuses solely on finding the final folded structure, which is already a very challenging problem. This approach assumes that the native structure of a protein corresponds to the global minimum of the potential energy of the protein. Therefore, the structure may be found directly, using optimization methods instead of dynamics simulation. However, global optimization is difficult, especially for functions with many local minima such as the potential energy functions for proteins. The success of this approach therefore depends on the development of a global optimization algorithm that can be applied effectively to proteins, motivating many intensive investigations along this line.

Selected Further Readings

Protein structure

Campbell NA, Reece JB, *Biology*, 7th ed., Benjamin Cummings, 2004.

Berg JM, Tymoczko JL, Stryer L, *Biochemistry*, W. H. Freeman, 2006.

Branden CI, Tooze J, *Introduction to Protein Structure*, 2nd ed., Garland Publishing Inc., 1999.

Lesk AM, *Introduction to Protein Architecture: The Structural Biology of Proteins*, Oxford University Press, 2001.

Creighton TE, *Proteins: Structures and Molecular Properties*, 2nd ed., Freeman & Co., 1993.

Jacob-Molina A, Arnold A, HIV reverse transcriptase function relationships, *Biochemistry* **30**: 6351–6361, 1991.

Rodgers DW, Gamblin SJ, Harris BA, Ray S, Culp JS, Hellmig B, Woolf DJ, Debouck C, Harrison SC, The structure of unliganded reverse transcriptase from the human immunodeficiency virus type 1, *Proc Natl Acad Sci USA* **92**: 1222–1226, 1995.

Telesnitsky A, Goff SP, Reverse transcriptase and the generation of retroviral DNA, in *Retroviruses*, Coffin J, Hughes S, Varmus H (eds.), Cold Spring Harbor Laboratory Press, pp. 121–160, 1997.

Zahn R, Liu A, Lührs T, Riek R, von Schroetter C, Garcia FL, Billeter M, Calzolai L, Wider G, Wüthrich K, NMR solution structure of the human prion protein, *Proc Natl Acad Sci USA* **97**: 145–150, 2000.

Aguzzi A, Montrasio F, Kaeser P, Prions: Health scare and biological challenge, *Nat Rev Mol Cell Biol* **2**: 118–125, 2001.

Aguzzi A, Heikenwalder M, Cannibals and garbage piles, *Nature* **423**: 127–129, 2003.

Structure determination

Woolfson MM, *Introduction to X-ray Crystallography*, 2nd ed., Cambridge University Press, 2003.
Drenth J, *Principles of Protein X-ray Crystallography*, 3rd ed., Springer, 2006.
Wuthrich K, *NMR of Proteins and Nucleic Acids*, Wiley, 1986.
Keeler J, *Understanding NMR Spectroscopy*, Wiley, 2005.
Schlick T, *Molecular Modelling and Simulation: An Interdisciplinary Guide*, Springer, 2003.
Bourne PE, Weissig H, *Structural Bioinformatics*, John Wiley & Sons, Inc., 2003.

Dynamics simulation

Haile JM, *Molecular Dynamics Simulation: Elementary Methods*, Wiley, 2001.
MaCammon JA, Harvey SA, *Dynamics of Proteins and Nucleic Acids*, Cambridge University Press, 2004.
Brooks III CL, Karplus M, Pettitt BM, *Proteins: A Theoretical Perspective of Dynamics, Structure, and Thermodynamics*, Advances in Chemical Physics, Vol. 71, Wiley, 2004.
Cui Q, Bahar I, *Normal Mode Analysis: Theory and Applications to Biological and Chemical Systems*, Chapman & Hall/CRC, 2005.

Protein folding

Dobson CM, Fersht AR, *Protein Folding*, Cambridge University Press, 1996.
Pain RH, *Mechanism of Protein Folding*, Oxford University Press, 2000.
Wolynes PG, Folding funnels and energy landscapes of larger proteins within the capillarity approximation, *Proc Natl Acad Sci USA* **94**: 6170–6175, 1997.
Mirny L, Shkhnovich E, Protein folding theory: From lattice to all-atom models, *Annu Rev Biophys Biomol Struct* **30**: 361–396, 2001.

Duan Y, Kollman P, Pathways to a protein folding intermediate observed in a 1-microsecond simulation in aqueous solution, *Science* **282**: 740–744, 1998.

Duan Y, Kollman PA, Computational protein folding: From lattice to all-atom, *IBM Syst J* **40**: 297–309, 2001.

Shirts MR, Pande VS, Screen savers of the world unite!, *Science* **290**: 1903–1904, 2000.

Pande VS, Baker I, Chapman J, Elmer SP, Khaliq S, Larson SM, Rhee YM, Shirts MR, Snow CD, Sorin EJ, Zagrovic B, Atomistic protein folding simulations on the submillisecond time scale using worldwide distributed computing, *Biopolymers* **68**: 91–109, 2002.

Elber R, Meller J, Olender R, Stochastic path approach to compute atomically detailed trajectories: Application to the folding of C peptide, *J Phys Chem* **103**: 899–911, 1999.

Wales DJ, Scheraga HA, Global optimization of clusters, crystals, and biomolecules, *Science* **285**: 1368–1372, 1999.

Chapter 2

X-ray Crystallography Computing

X-ray crystallography is a major experimental tool for structure determination. In order to use this technique, a protein sample has to be gathered, purified, and crystallized, which usually takes at least several months of lab work. The protein crystal is then applied with X-ray to obtain a diffraction image. The image shows only the diffraction intensities and patterns, not the three-dimensional (3D) structure, of the protein. The latter must be further derived from the obtained diffraction image, requiring the solution of a well-known problem called the phase problem.

2.1. The Phase Problem

The phase problem has been an intensive research target for more than half a century in the field of X-ray crystallography. A crystal structure cannot be fully revealed without the solution of the phase problem, either heuristically with the insights from various theoretical or experimental sources or directly using a mathematical or computational approach. In either case, a reliable and accurate solution to the problem has generally proven to be difficult to achieve.

2.1.1. *X-ray diffraction*

Molecules in the crystalline state form regular patterns, with the atoms appearing periodically in space. The electrons around the atoms, when incident to an X-ray beam, become energetically excited and start scattering their own X-rays. Due to the periodic distribution of the atoms in the crystal, the scattered X-rays enhance in a discrete set of directions and produce so-called X-ray diffractions or reflections. The latter can be recorded by using a diffraction detector, and the patterns and intensities of the diffractions can then be based upon to deduce the electron density distribution in the crystal and hence the structure of the protein (see Fig. 2.1).

The configuration of the electrons, which move rapidly around the atoms, can only be described by electron density distribution. Once the electron density distribution is known, the atoms can be identified in regions with high electron densities. The X-ray diffractions, when projected onto a two-dimensional (2D) plane, form a lattice of light spots, with each spot corresponding to one diffraction beam. The characteristics of each diffracted light can be described by a complex value called the structure factor. The magnitude (or the amplitude) of the complex value is correlated with the intensity of the corresponding light spot, which can therefore be measured from the diffraction image. However, the complex argument, also called the phase, cannot be detected from the experiments. Otherwise,

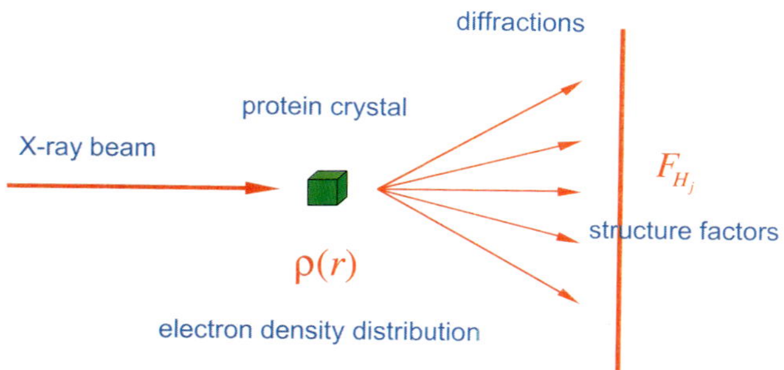

Fig. 2.1. X-ray crystallography. The electrons in the crystal are excited when incident to the incoming X-rays and start scattering X-rays. The scattered X-rays from different electrons enhance in certain directions and form the X-ray diffractions.

knowledge of both the magnitudes and arguments of the structure factors would suffice for the full determination of the electron density distribution.

Let $\rho(r)$ be the electron density distribution function for a crystal, where r is the position vector in 3D space. Let F_H be a structure factor, where H is a 3D integer vector serving as the index for the factor. Then, the correlation between the electron density distribution and the X-ray diffractions can be described rigorously in terms of a mathematical relationship between the electron density distribution function and the structure factors that describe the diffractions:

$$\rho(r) = \sum_{H \in Z^3} F_H \exp(-2\pi i H \cdot r), \tag{2.1}$$

$$F_H = \int_V \rho(r) \exp(2\pi i H \cdot r) dr, \quad H \in Z^3, \tag{2.2}$$

where V is the volume of a unit cell of the crystal and $H \cdot r$ is the inner product of H and r.

Formulas (2.1) and (2.2) show that the electron density distribution function can be expressed as a Fourier series with the structure factors as the coefficients, while each structure factor is a Fourier transform of the electron density distribution function. Clearly, the electron density distribution and hence the structure of the crystal can be obtained if all of the structure factors of the crystal are known, and vice versa. Unfortunately, the structure factors can only be detected partially in the diffraction experiments, with their complex arguments (phases) yet to be determined. The latter gives rise to the famous phase problem in X-ray crystallography.

2.1.2. *Electron scattering*

As a light wave, an X-ray travels in space, and its electromagnetic strength behaves as a cosine function of position z along a straight line. Let us use X to represent an X-ray beam as well as its electromagnetic strength. Then,

$$X = A \cos \frac{2\pi}{\lambda}(z + \phi), \tag{2.3}$$

where A is the amplitude; λ is the wavelength; and ϕ is the initial position or, more accurately, the initial phase.

Let X_1 and X_2 be two X-rays scattered from electrons e_1 and e_2 when they are incident to an incoming X-ray X, where e_1 and e_2 are located at p_1 and p_2, respectively (Fig. 2.2). Then, X_1 and X_2 have the same wavelength as X, and

$$X_1 = A_1 \cos \frac{2\pi}{\lambda}(z + \phi_1),$$
$$X_2 = A_2 \cos \frac{2\pi}{\lambda}(z + \phi_2),$$

(2.4)

where ϕ_1 and ϕ_2 can be different, and A_1 and A_2 depend on the electron densities of the scattering sources.

The electrons e_1 and e_2 both scatter X-rays in all directions, but only those in the same direction may add up and make diffractions. So, assume that X travels in direction s, and consider X_1 and X_2 in direction s', and $||s|| = ||s'|| = 1$, where $|| \cdot ||$ is the Euclidean norm and represents the length of a vector. Then, at a position z, X_2 must have traveled a longer distance or, in other words, started earlier than X_1 by the amount of

$$z + \phi_2 - (z + \phi_1) = d + d' = r_2 \cdot s - r_2 \cdot s' = r_2 \cdot S,$$

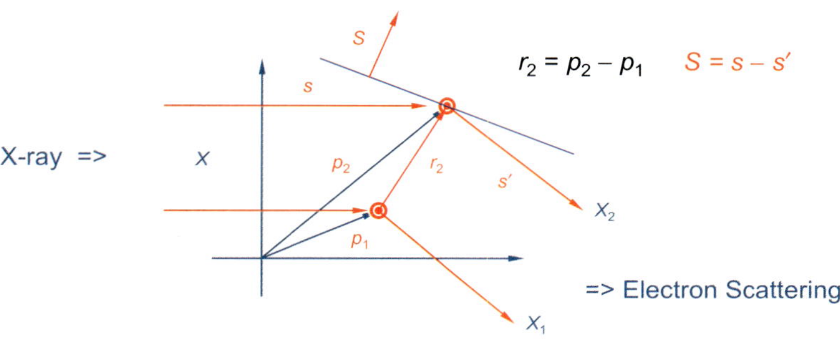

Fig. 2.2. X-ray scattering. The electrons at positions p_1 and p_2 are incident to X-ray X and start scattering X-rays. The scattered X-rays X_1 and X_2 in the same direction add to become a diffraction.

where $r_2 = p_2 - p_1$ and $S = s - s'$. Consider the case when e_2 is located at the origin with $p_2 = 0$ and $\phi_2 = 0$. We then obtain

$$\phi_1 = p_1 \cdot s - p_1 \cdot s' = p_1 \cdot S,$$
$$\phi_2 = r_2 \cdot S + p_1 \cdot S = p_2 \cdot S.$$

We now consider how to add X_1 and X_2. From formula (2.3), X can be expressed in the following form:

$$X = A \cos\left(\frac{2\pi}{\lambda}\phi\right) \cos\left(\frac{2\pi}{\lambda}z\right) + A \sin\left(\frac{2\pi}{\lambda}\phi\right) \cos\left(\frac{2\pi}{\lambda}z + \frac{\pi}{2}\right).$$

Since the two cosine functions, $\cos\left(\frac{2\pi}{\lambda}z\right)$ and $\cos\left(\frac{2\pi}{\lambda}z + \frac{\pi}{2}\right)$, are linearly independent, X can be considered as a combination of two independent components defined by the two functions. The amount along the first and second components are determined by $A \cos\left(\frac{2\pi}{\lambda}\phi\right)$ and $A \sin\left(\frac{2\pi}{\lambda}\phi\right)$, respectively. Therefore, X can be represented more compactly in a complex expression,

$$X = A\left(\cos\frac{2\pi}{\lambda}\phi + i\sin\frac{2\pi}{\lambda}\phi\right) = A \exp\left(\frac{2\pi i}{\lambda}\phi\right), \qquad (2.5)$$

and X_1 and X_2 can be written as

$$X_1 = A_1 \exp\left(\frac{2\pi i}{\lambda}\phi_1\right),$$
$$X_2 = A_2 \exp\left(\frac{2\pi i}{\lambda}\phi_2\right). \qquad (2.6)$$

With these complex representations, adding X_1 and X_2 becomes simply adding two complex numbers:

$$X_1 + X_2 = A_1 \exp\left(\frac{2\pi i}{\lambda}\phi_1\right) + A_2 \exp\left(\frac{2\pi i}{\lambda}\phi_2\right). \qquad (2.7)$$

Since $\phi_1 = p_1 \cdot S$ and $\phi_2 = p_2 \cdot S$, the addition can also be written as

$$X_1 + X_2 = A_1 \exp\left(\frac{2\pi i}{\lambda}p_1 \cdot S\right) + A_2 \exp\left(\frac{2\pi i}{\lambda}p_2 \cdot S\right) \qquad (2.8)$$

or, when necessary, as

$$X_1 + X_2 = F \exp\left(\frac{2\pi i}{\lambda} p_1 \cdot S\right), \tag{2.9}$$

with

$$F = A_1 + A_2 \exp\left(\frac{2\pi i}{\lambda} r_2 \cdot S\right),$$

where F is equivalent to the sum of X_1 and X_2 when e_1 is assumed to be at the origin.

In general, if there are n electrons, $e_1, e_2, \ldots, e_n$, at positions $p_1, p_2, \ldots, p_n$, let $r_j = p_j - p_1$, $j = 1, 2, \ldots, n$. Then, we can have the following general formulas for adding all of the scattered X-rays from these electrons:

$$\sum_{j=1}^{n} X_j = \sum_{j=1}^{n} A_j \exp\left(\frac{2\pi i}{\lambda} p_j \cdot S\right) \tag{2.10}$$

or

$$\sum_{j=1}^{n} X_j = F \exp\left(\frac{2\pi i}{\lambda} p_1 \cdot S\right), \tag{2.11}$$

with

$$F = \sum_{j=1}^{n} A_j \exp\left(\frac{2\pi i}{\lambda} r_j \cdot S\right).$$

Note that for any diffraction direction s', we can always find a plane T so that the diffraction in direction s' can be regarded as a reflection on T from the incoming X-ray in direction s. The reflection angle θ is equal to half of the angle between s and s', and S is perpendicular to T. Because of this property, X-ray diffraction is sometimes also called X-ray reflection. It can be specified by either its diffraction direction s' or the normal vector S of the reflection plane.

2.1.3. *Atomic scattering factor*

We now consider the X-rays that can be scattered from an entire atom. Each atom has a group of electrons moving and in fact forming an electron cloud around the nucleus. When the atom is incident to an X-ray beam, all of its electrons will contribute to scattering new X-rays. Assume that the electron cloud has a symmetric shape and that the scattered X-rays add up as if they form single X-rays scattered from the center of the atom. Let the center of the atom be located at q. Let X_p be the X-ray scattered from a position p around q (Fig. 2.3). Then,

$$X_p = \rho(p - q) \exp\left(\frac{2\pi i}{\lambda} \phi_p\right),$$

where ρ is the electron density distribution function for the atom. Let $S = s - s'$ and $r = p - q$. Then, $\phi_q = q \cdot S$, $\phi_p = p \cdot S$, and $\phi_p - \phi_q = r \cdot S$. Let Y be the sum of the X-rays scattered from the entire atom in direction s'. Then, similar to formulas (2.10) and (2.11) for a group of electrons, we have for a continuous electron cloud

$$\begin{aligned}
Y &= \int_{p \in R^3} X_p \, dp \\
&= \int_{p \in R^3} \rho(p - q) \exp\left(\frac{2\pi i}{\lambda} p \cdot S\right) dp \\
&= \exp\left(\frac{2\pi i}{\lambda} q \cdot S\right) \int_{r \in R^3} \rho(r) \exp\left(\frac{2\pi i}{\lambda} r \cdot S\right) dr.
\end{aligned}$$

In a shorter form,

$$Y = f \exp\left(\frac{2\pi i}{\lambda} q \cdot S\right), \tag{2.12}$$

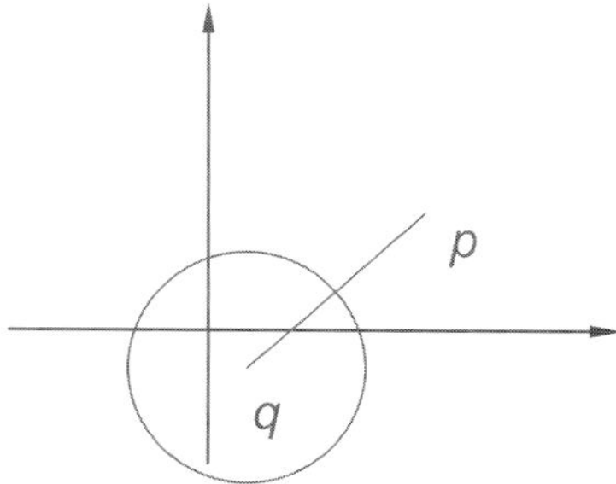

Fig. 2.3. Atomic scattering. The diffraction from an atom at q is the sum of the diffractions from all of the electrons at p around q.

with

$$f = \int_{r \in R^3} \rho(r) \exp\left(\frac{2\pi i}{\lambda} r \cdot S\right) dr.$$

Here, since ρ is symmetric, the integral is zero for the imaginary part of the integrand. Therefore, f can be evaluated as

$$f = \int_{r \in R^3} \rho(r) \cos\left(\frac{2\pi}{\lambda} r \cdot S\right) dr, \tag{2.13}$$

and it is in general real. The value of f depends only on the electron density distribution around the atom of interest. It shows the X-ray scattering capacity for the atom and is therefore called the atomic scattering factor.

2.1.4. *Crystal lattice*

A unit cell is the smallest volume that repeats in the crystal space, and it is usually the volume that contains one single molecule. Suppose that in each unit cell, there are n atoms. Then, the X-rays scattered from the unit cell can be calculated as the sum of those from all atoms in the cell. Let the center of atom j be located at p_j, $j = 1, 2, \ldots, n$. Let Y_j be the X-ray scattered from atom j in direction s'. Then, by formula (2.12),

$$Y_j = f_j \exp\left(\frac{2\pi i}{\lambda} p_j \cdot S\right), \tag{2.14}$$

with

$$f_j = \int_{r \in R^3} \rho_j(r) \cos\left(\frac{2\pi}{\lambda} r \cdot S\right) dr,$$

where f_j and ρ_j are the atomic scattering factor and the electron density distribution function, respectively, for atom j, $j = 1, 2, \ldots, n$.

Let Z be the X-ray scattered from the entire unit cell. Then,

$$Z = \sum_{j=1}^{n} Y_j = \sum_{j=1}^{n} f_j \exp\left(\frac{2\pi i}{\lambda} p_j \cdot S\right)$$

$$= \exp\left(\frac{2\pi i}{\lambda} p_1 \cdot S\right) \sum_{j=1}^{n} f_j \exp\left(\frac{2\pi i}{\lambda} r_j \cdot S\right),$$

where $r_j = p_j - p_1$, $j = 1, 2, \ldots, n$. In a shorter form,

$$Z = F_S \exp\left(\frac{2\pi i}{\lambda} p_1 \cdot S\right), \tag{2.15}$$

with

$$F_S = \sum_{j=1}^{n} f_j \exp\left(\frac{2\pi i}{\lambda} r_j \cdot S\right).$$

Here, F_S is called the structure factor of a unit cell of the crystal.

Suppose that a unit cell is a parallelogram extended by three vectors a, b, c (Fig. 2.4). The cell repeats in space to form a crystal lattice:

$$L = \{ua + vb + wc : u, v, w \in Z\}. \tag{2.16}$$

Suppose that $0 \leq u \leq n_1 - 1$, $0 \leq v \leq n_2 - 1$, and $0 \leq w \leq n_3 - 1$. Then, there is a unit cell at each lattice point $p = ua + vb + wc$ and there are a total of $n_1 \times n_2 \times n_3$ unit cells in the crystal. Assume that there are n atoms in each of the unit cells with positions at $p = p_1, p_2, \ldots, p_n$. Then, for a cell at p, we have

$$Z_p = F_S \exp\left(\frac{2\pi i}{\lambda} p \cdot S\right), \tag{2.17}$$

with

$$F_S = \sum_{j=1}^{n} f_j \exp\left(\frac{2\pi i}{\lambda} r_j \cdot S\right),$$

and the X-ray scattered from the entire crystal in direction S can be calculated as the sum of those from all of the unit cells:

$$K_S = \sum_{p \in L} Z_p = F_S \sum_{p \in L} \exp\left(\frac{2\pi i}{\lambda} p \cdot S\right).$$

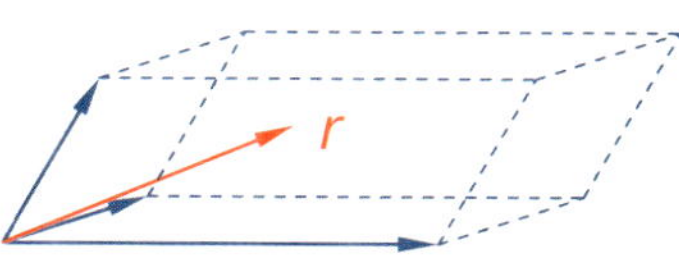

Fig. 2.4. Crystal unit. If the unit is formed by three vectors a, b, c, any point in the unit $r = xa + yb + zc$.

Note that the sum on the right-hand side can be expanded to

$$\sum_{u=0}^{n_1-1} \exp\left(\frac{2\pi i}{\lambda} ua \cdot S\right) \sum_{v=0}^{n_2-1} \exp\left(\frac{2\pi i}{\lambda} vb \cdot S\right) \sum_{w=0}^{n_3-1} \exp\left(\frac{2\pi i}{\lambda} wc \cdot S\right).$$

Since n_1, n_2, and n_3 are all large numbers, the summations will be equal to zero unless $a \cdot S, b \cdot S$, and $c \cdot S$ are integer multiples of λ. In other words, the crystal will produce diffractions only if

$$a \cdot S = \lambda h$$
$$b \cdot S = \lambda k \qquad\qquad (2.18)$$
$$c \cdot S = \lambda l$$

for some integers h, k, and l. These conditions are called the Laue conditions of X-ray crystallography.

Let V be the volume of a unit cell. We can write the structure factor F_S as an integral for all of the electrons in V:

$$F_S = \int_V \rho(r) \exp\left(\frac{2\pi i}{\lambda} r \cdot S\right) dr,$$

where ρ is the electron density distribution in the entire unit cell V.

Let $r = xa + yb + zc$. In order for F_S to be nonzero, S must satisfy the Laue conditions. Therefore,

$$F_S = \int_V \rho(r) \exp\left(\frac{2\pi i}{\lambda} r \cdot S\right) dr$$
$$= \int_V \rho(r) \exp\left[\frac{2\pi i}{\lambda}(xa \cdot S + yb \cdot S + zc \cdot S)\right] dr$$
$$= \int_V \rho(r) \exp\left[2\pi i(hx + ky + lz)\right] dr.$$

Let $H = (h, k, l)^\mathrm{T}$ and $r = (x, y, z)^\mathrm{T}$. Let F_S be replaced by F_H. We then have

$$F_H = \int_V \rho(r) \exp\left(2\pi i H \cdot r\right) dr, \quad H \in Z^3. \tag{2.19}$$

We now express ρ as a Fourier series,

$$\rho(r) = \sum_{K \in Z^3} A_K \exp(-2\pi i K \cdot r),$$

and substitute it into formula (2.19). We then obtain

$$F_H = \int_V \left\{ \sum_{K \in Z^3} A_K \exp\left[2\pi i (H - K) \cdot r\right] \right\} dr$$

$$= \sum_{K \in Z^3} A_K \int_V \exp\left[2\pi i (H - K) \cdot r\right] dr = A_H, \quad H \in Z^3,$$

and therefore,

$$\rho(r) = \sum_{H \in Z^3} F_H \exp(-2\pi i H \cdot r). \tag{2.20}$$

2.2. Least Squares Solutions

It is difficult to find the phases for the formation of a correct crystal structure given only the amplitudes of the structure factors. First, there are infinitely many phases to be determined, while only a finite number of diffraction spots can be measured. Therefore, only a finite number of structure factors can eventually be determined, with which an approximate crystal structure can be constructed. The quality of the structure varies with the availability and accuracy of the diffraction data. Second, even with a commonly available set of diffraction data, say for around 10 000 structure factors, the determination of the phases presents a great computational challenge because a large nonlinear system of equations with around 10 000 variables needs to be solved. An accurate solution to the system has proven difficult to obtain if without a genuinely designed algorithm.

2.2.1. *Diffraction equations*

In terms of atomic scattering factors, the structure factors can be written in the following form:

$$F_H = \sum_{j=1}^{n} f_j \exp(2\pi i H \cdot r_j), \quad H \in Z^3,$$

where f_j and r_j are the atomic scattering factor and the position vector of atom j, respectively. Because F_H is a complex number, $F_H = |F_H| \exp(i\phi_H)$, where the magnitude $|F_H|$ and complex argument ϕ_H are also called the amplitude and phase of F_H, respectively. It follows that

$$|F_H| \exp(i\phi_H) = \sum_{j=1}^{n} f_j \exp(2\pi i H \cdot r_j), \quad H \in Z^3. \tag{2.21}$$

Each of the above equations is a complex equation and can be split into two real equations:

$$|F_H| \cos \phi_H = \sum_{j=1}^{n} f_j \cos(2\pi H \cdot r_j),$$

$$|F_H| \sin \phi_H = \sum_{j=1}^{n} f_j \sin(2\pi H \cdot r_j). \tag{2.22}$$

If m structure factors are involved, there will be $2m$ such equations. If the phases ϕ_H and the atomic positions r_j in the equations are considered as unknowns while the amplitudes $|F_H|$ are as given, then there will be $m + n$ unknowns in total. Therefore, in principle, we may be able to solve the equations for the phases and the atomic positions simultaneously as long as we have a greater number of equations than unknowns, i.e. $m \gg n$.

2.2.2. Sayre's equations

For practical purposes, structure factors are sometimes scaled or normalized. A unitary structure factor, usually denoted as U_H, is defined as the structure factor F_H divided by the sum of the atomic scattering factors, i.e.

$$U_H = \frac{F_H}{\sum_{j=1}^{n} f_j} = \sum_{j=1}^{n} n_j \exp(2\pi i H \cdot r_j), \qquad (2.23)$$

where

$$n_j = \frac{f_j}{\sum_{j=1}^{n} f_j}.$$

A normalized structure factor, usually denoted as E_H, is defined as the structure factor F_H normalized by the square root of the sum of squares of the atomic scattering factors, i.e.

$$E_H = \frac{F_H}{\left(\sum_{j=1}^{n} f_j^2\right)^{1/2}} = \sum_{j=1}^{n} z_j \exp(2\pi i H \cdot r_j), \qquad (2.24)$$

where

$$z_j = \frac{f_j}{\left(\sum_{j=1}^{n} f_j^2\right)^{1/2}}.$$

Let F_H be the structure factor for the density function ρ and Z_H the structure factor for ρ^2. Assume that the atoms in the molecule are all equal and there is no overlap among the atoms. Then,

$$F_H = f \sum_{j=1}^{n} \exp(2\pi i H \cdot r_j),$$

where f is the atomic scattering factor calculated with the density distribution function ρ; and

$$Z_H = f^* \sum_{j=1}^{n} \exp(2\pi i H \cdot r_j),$$

where f^* is the atomic scattering factor calculated with the density distribution function ρ^2. It follows that F_H is proportional to Z_H, i.e.

$$F_H = C Z_H, \quad H \in Z^3,\tag{2.25}$$

where C is a constant, $C = f / f^*$.

By the definition of Z_H,

$$Z_H = \int_V \rho^2(r) \exp\left(-2\pi i H \cdot r\right) dr$$

$$= \int_V \left[\sum_K F_K \exp(2\pi i K \cdot r)\right] \rho(r) \exp(-2\pi i H \cdot r) dr$$

$$= \sum_K F_K \int_V \rho(r) \exp\left[-2\pi i (H - K) \cdot r\right] dr$$

$$= \sum_K F_K F_{H-K},$$

and therefore Z_H is in fact a self-convolution of F_H, i.e.

$$Z_H = \sum_K F_K F_{H-K}, \quad H \in Z^3.\tag{2.26}$$

It follows that

$$F_H = C \sum_K F_K F_{H-K}, \quad H \in Z^3.\tag{2.27}$$

The above equations show some important relations among structure factors, which are often used as additional information for phase determination. The equations are called Sayre's equations because of Sayre's early work on these relations. They can also be written in other equivalent forms such as

$$|F_H|^2 = C \sum_K F_H F_K F_{-H-K}, \quad H \in Z^3,\tag{2.28}$$

$$|U_H|^2 = C \sum_K U_H U_K U_{-H-K}, \quad H \in Z^3,\tag{2.29}$$

$$|E_H|^2 = C \sum_K E_H E_K E_{-H-K}, \quad H \in Z^3.\tag{2.30}$$

2.2.3. *Cochran distributions*

In terms of atomic scattering factors, a unitary structure factor U_H can be written in the following form:

$$U_H = \sum_{j=1}^{n} n_j \exp(2\pi i H \cdot r_j)$$

$$= \sum_{j=1}^{n} n_j \exp(2\pi i K \cdot r_j) \exp\left[2\pi i (H - K) \cdot r_j\right].$$

Then, by averaging both sides of the equation over H and K, we have

$$\langle U_H \rangle = \sum_{j=1}^{n} n_j \langle \cos(2\pi K \cdot r_j) \cos[2\pi(H - K) \cdot r_j]\rangle$$

$$= \langle \cos(2\pi K \cdot r_j) \cos[2\pi(H - K) \cdot r_j]\rangle.$$

Note that

$$U_K U_{H-K} = \sum_{j=1}^{n}\sum_{k=1}^{n} n_j n_k \exp(2\pi i K \cdot r_j) \exp\left[2\pi i (H - K) \cdot r_k\right].$$

Then,

$$U_K U_{H-K} = \sum_{j=1}^{n}\sum_{k=1}^{n} n_j n_k \cos(2\pi K \cdot r_j) \cos[2\pi(H - K) \cdot r_k]$$

$$\approx \langle \cos(2\pi K \cdot r_j) \cos[2\pi(H - K) \cdot r_k]\rangle,$$

and approximately,

$$\langle U_H \rangle = U_K U_{H-K}. \tag{2.31}$$

Applying this relation to the amplitude and the phase separately, we have

$$\langle |U_H| \rangle = |U_K U_{H-K}|,$$

$$\langle \phi_H \rangle = \phi_K + \phi_{H-K}. \tag{2.32}$$

Now, let $U_H = A_H + iB_H$, and

$$x = A_H - \langle A_H \rangle,$$
$$y = B_H - \langle B_H \rangle.$$

If all of the atoms are equal and $\langle A_H \rangle$ and $\langle B_H \rangle$ are assumed to be zero, then

$$\langle x^2 \rangle = \langle A_H^2 \rangle - \langle A_H \rangle^2$$
$$= \sum_{j=1}^{n} n_j^2 \left[\langle \cos^2(2\pi H \cdot r_j) \rangle - \langle \cos(2\pi H \cdot r_j) \rangle^2 \right]$$
$$= \sum_{j=1}^{n} n_j^2 \left[\frac{1}{2} + \frac{1}{2} \langle \cos(4\pi H \cdot r_j) \rangle - \langle \cos(2\pi H \cdot r_j) \rangle^2 \right]$$
$$= \frac{1}{2n}.$$

Following the same argument, we also have

$$\langle y^2 \rangle = \langle B_H^2 \rangle - \langle B_H \rangle^2$$
$$= \sum_{j=1}^{n} n_j^2 \left(\langle \sin^2(2\pi H \cdot r_j) \rangle - \langle \sin(2\pi H \cdot r_j) \rangle^2 \right)$$
$$= \sum_{j=1}^{n} n_j^2 \left(\frac{1}{2} - \frac{1}{2} \langle \cos(4\pi H \cdot r_j) \rangle - \langle \sin(2\pi H \cdot r_j) \rangle^2 \right)$$
$$= \frac{1}{2n}.$$

Let σ be the standard deviation for (A_H, B_H). Then,

$$\sigma^2 = \langle x^2 + y^2 \rangle = \frac{1}{n}. \tag{2.33}$$

By the central limit theorem, the joint probability of A_H and B_H is then approximately equal to a normal distribution:

$$P(A_H, B_H) = \frac{n}{\pi} \exp\left[-n(x^2 + y^2)\right]. \tag{2.34}$$

Because $\tan \phi_H = B_H/A_H$, the probability of ϕ_H must be proportional to $P(A_H, B_H)$, i.e.

$$P(\phi_H) = cP(A_H, B_H),$$

where c is a constant. Since

$$\int_0^{2\pi} P(\phi_H)\, d\phi_H = 1,$$

we then have the value for the constant c:

$$c = \frac{1}{\int_0^{2\pi} P(A_H, B_H)\, d\phi_H}.$$

We now write

$$\begin{aligned}
x^2 + y^2 &= |A_H|^2 - 2A_H \cdot \langle A_H \rangle + |\langle A_H \rangle|^2 \\
&\quad + |B_H|^2 - 2B_H \cdot \langle B_H \rangle + |\langle B_H \rangle|^2 \\
&= |U_H|^2 - 2U_H \cdot \langle U_H \rangle + |\langle U_H \rangle|^2.
\end{aligned}$$

Note that

$$\langle U_H \rangle = U_K U_{H-K},$$
$$U_H \cdot \langle U_H \rangle = |U_H||U_K U_{H-K}| \cos(\phi_H - \langle \phi_H \rangle).$$

It follows that

$$x^2 + y^2 = |U_H|^2 + |U_K U_{H-K}|^2 - 2|U_H U_K U_{H-K}| \cos(\phi_H - \langle \phi_H \rangle)$$

and

$$P(\phi_H) = \frac{\exp(A_{HK} \cos \phi_{HK})}{\int_0^{2\pi} \exp(A_{HK} \cos \phi_{HK})\, d\phi_H}, \qquad (2.35)$$

where $A_{HK} = 2|U_H U_K U_{H-K}|$ and $\phi_{HK} = \phi_H - \phi_K - \phi_{H-K}$. Given the facts that $P(\phi_H) = P(\phi_{HK})$ and $d\phi_H = d\phi_{HK}$ for fixed K, we can also write the above formula in a more general form:

$$P(\phi_{HK}) = \frac{\exp(A_{HK} \cos \phi_{HK})}{\int_0^{2\pi} \exp(A_{HK} \cos \phi_{HK})\, d\phi_{HK}}. \qquad (2.36)$$

Formulas (2.35) and (2.36) are known as the Cochran distributions of the phases, and are important statistical properties that can be used for phase evaluations.

2.2.4. *Minimal principles*

Various least squares formulas can be defined for phase determination or refinement using the general equation (2.21) or (2.22), or with other phase relations such as one of the Sayre's equations (2.27)–(2.30) or the Cochran distribution (2.35) or (2.36).

For the diffraction equation (2.21) or (2.22), a general solution principle is to formulate the equations into a least squares problem so that the phases and atomic positions can be found to minimize all of the violations of the equations. Mathematically, the problem is to find a set of values for the phases ϕ_H and the atomic positions r_j so that the sum of squared errors of the equations is minimized, i.e.

$$\min_{\phi,r} \; F_{\phi,r}^2 + G_{\phi,r}^2, \qquad (2.37)$$

with

$$F_{\phi,r}^2 = \sum_{H \in S} \left[|F_H| \cos(\phi_H) - \sum_{j=1}^{n} f_j \cos(2\pi H \cdot r_j) \right]^2 ,$$

$$G_{\phi,r}^2 = \sum_{H \in S} \left[|F_H| \sin(\phi_H) - \sum_{j=1}^{n} f_j \sin(2\pi H \cdot r_j) \right]^2 ,$$

where $\phi = \{\phi_H : H \text{ in } S \subset Z^3\}$ and $r = \{r_j : j = 1, 2, \ldots, n\}$. This problem can be solved by using a conventional optimization method. However, the objective function usually has many local minima and the global minimum is hard to obtain. In practice, a local minimum may be used as an approximate solution to the problem when the objective function is minimized to a sufficiently small value. Even such a local minimum may not necessarily be easy to find if the initial solution is not good enough.

An alternative approach is to utilize additional information from one of the Sayre's equations (2.27)–(2.30) so that the phases are determined to satisfy not only the general diffraction equations, but also the phase relations in Sayre's equations. Let us consider Eq. (2.30) and write it in the following equivalent form:

$$|E_H|^2 = C \sum_{K} E_H E_{-K} E_{K-H}, \quad H \in Z^3.$$

Then, by taking the real and imaginary parts of the equation, we have

$$|E_H|^2 = C \sum_K |E_H E_K E_{H-K}| \cos(\phi_H - \phi_K - \phi_{H-K}),$$

$$0 = C \sum_K |E_H E_K E_{H-K}| \sin(\phi_H - \phi_K - \phi_{H-K}), \quad H \in Z^3.$$

Let $A_{HK} = |U_H U_K U_{H-K}|$ and $\phi_{HK} = \phi_H - \phi_K - \phi_{H-K}$. Then, the equations become

$$|E_H|^2 = C \sum_K A_{HK} \cos \phi_{HK},$$

$$0 = C \sum_K A_{HK} \sin \phi_{HK}, \qquad H \in Z^3.$$

By using these equations, another least squares problem can immediately be obtained for the determination of the phases:

$$\min_{\phi,r} \sum_{\phi_{HK}} A^2_{\phi_{HK}} + B^2_{\phi_{HK}}, \tag{2.38}$$

with

$$A_{\phi_{HK}} = |E_H|^2 - C \sum_K A_{HK} \cos \phi_{HK},$$

$$B_{\phi_{HK}} = \sum_K A_{HK} \sin \phi_{HK},$$

where $\phi = \{\phi_H : H \text{ in } S \subset Z^3\}$ and $r = \{r_j : j = 1, 2, \ldots, n\}$ should also be consistent with the general diffraction equation (2.21) or (2.22).

Using the Cochran distribution (2.35) or (2.36), another set of phase refinement formulas can be obtained as well. For example, the expectation of $\cos \phi_{HK}$ can be evaluated by the formula

$$\langle \cos \phi_{HK} \rangle = \frac{\int_0^{2\pi} \cos \phi_{HK} \exp\left(A_{HK} \cos \phi_{HK}\right) d\phi_{HK}}{\int_0^{2\pi} \exp\left(A_{HK} \cos \phi_{HK}\right) d\phi_{HK}}. \tag{2.39}$$

With this property, a least squares problem can be formulated so that the phases can be determined to make the cosines of the phase triplets,

ϕ_{HK}, to be close to their expectation values as much as possible, i.e.

$$\min_{\phi,r} R_{\phi,r} = \frac{\sum_{HK} A_{HK}[\cos \phi_{HK} - \langle \cos \phi_{HK} \rangle]^2}{\sum_{HK} A_{HK}}, \tag{2.40}$$

where A_{HK} are used as the weights for the errors, and $\phi = \{\phi_H : H \text{ in } S \subset Z^3\}$ and $r = \{r_j : j = 1, 2, \ldots, n\}$ should again be consistent with the general diffraction equation (2.21) or (2.22).

2.3. Entropy Maximization

Let the unit cell of a crystal be divided into n boxes, and let p_j be the percentage of total electrons in box j and $\Sigma_j p_j = 1$. Then, the entropy of the crystal can be defined as

$$S_0(p) = -\sum_{j=1}^{n} p_j \log p_j, \tag{2.41}$$

where $p = (p_1, p_2, \ldots, p_n)^{\mathrm{T}}$. Let q_j be some prior probability of having an electron in box j. Then, a relative entropy S can be defined such that

$$S(p) = -\sum_{j=1}^{n} p_j \log \left(\frac{p_j}{q_j} \right). \tag{2.42}$$

Here, $S(p) = 0$ if $p_j = q_j$ for all j, and $S(p) < 0$ otherwise.

In general, for a crystal system with a continuously defined electron density distribution function ρ, the entropy of the system can be evaluated using the formula

$$S_0(\rho) = -\int_V \rho(r) \log \rho(r) \, dr, \tag{2.43}$$

where V is the volume of the unit cell of the crystal. Similarly, with respect to some prior probability distribution q, a relative entropy S can be defined such that

$$S(\rho) = -\int_V \rho(r) \log \left[\frac{\rho(r)}{q(r)} \right] dr. \tag{2.44}$$

Here, $S(\rho) = 0$ if $\rho = q$, and $S(\rho) < 0$ otherwise.

2.3.1. *Entropy vs. probability*

Let us first consider the discrete case when the unit cell is divided into n boxes. Suppose that there are N electrons. If there are n_j electrons in box j, there will be in total $N!/(n_1!n_2!\cdots n_n!)$ arrangements of the electrons in the boxes. If the prior probability of putting an electron in box j is q_j, the probability of having one of the above arrangements of the electrons will be

$$P(n_1, n_2, \ldots, n_n) = \frac{N!}{n_1!n_2!\cdots n_n!} q_1^{n_1} q_2^{n_2} \cdots q_n^{n_n}. \qquad (2.45)$$

With Stirling's formula, the probability can be calculated approximately as

$$\log P(n_1, n_2, \ldots, n_n)$$
$$\approx N \log N - N - \sum_{j=1}^{n}(n_j \log n_j - n_j) + \sum_{j=1}^{n} n_j \log q_j$$
$$= -N \sum_{j=1}^{n} \left(\frac{n_j}{N}\right) \log \left[\frac{(n_j/N)}{q_j}\right].$$

Let $p_j = n_j/N$, $j = 1, 2, \ldots, n$. Then, p_j is exactly the electron density in box j, and the probability of the electron arrangement $P(n_1, n_2, \ldots, n_n)$ is exactly the probability of the electron density distribution $P(p_1, p_2, \ldots, p_n)$. Let $p = (p_1, p_2, \ldots, p_n)^{\mathrm{T}}$. Then,

$$\log P(p) = -N \sum_{j=1}^{n} p_j \log \left(\frac{p_j}{q_j}\right) = NS(p), \qquad (2.46)$$

where S is the entropy function defined in Eq. (2.42). It follows that

$$P(p) = \exp[NS(p)].$$

The above formulas can be generalized to the continuous case so that for any continuously defined electron density distribution function ρ,

$$\log P(\rho) = -N \int_V \rho(r) \log \left[\frac{\rho(r)}{q(r)}\right] dr = NS(\rho), \qquad (2.47)$$

where S is the entropy function defined in Eq. (2.44), and

$$P(\rho) = \exp[NS(\rho)].$$

From Eqs. (2.46) and (2.47), we see that the entropy for a crystal system with an electron density distribution ρ is correlated with the probability of ρ, and the maximum entropy $S(\rho)$ corresponds to the highest probability $P(\rho)$.

2.3.2. *Maximizing entropy*

The entropy maximization approach to the phase problem determines the phases based on the evaluation and, in particular, the maximization of the entropy defined in terms of the corresponding electron density. Central to this approach is a constrained entropy maximization problem, which maximizes the entropy of a given crystal system subject to a set of constraints requiring the structure factors to be equal to some given values. Formally, the problem can be stated as

$$\max_{\rho} S(\rho) \tag{2.48}$$

with

$$\int_V \rho(r) \exp\left(-2\pi i H_j \cdot r\right) dr = F^*_{H_j}, \quad j = 1, 2, \ldots, m,$$

$$\int_V \rho(r)\, dr = 1,$$

where $F^*_{H_j}$ are given structure factors with known amplitudes and predicted phases, and

$$S(\rho) = -\int_V \rho(r) \log\left[\frac{p(r)}{q(r)}\right] dr.$$

For convenience, we write the problem in the following more general form:

$$\max_{\rho} S(\rho) \tag{2.49}$$

with

$$F_{H_j}(\rho) = F_{H_j}^*, \quad j = 1, 2, \ldots, m,$$
$$F_{H_0}(\rho) = F_{H_0}^* = 1,$$

where

$$F_{H_j}(\rho) = \int_V \rho(r) C_{H_j}(r)\, dr = \langle C_{H_j} \rangle, \quad j = 0, 1, \ldots, m$$

with

$$C_{H_j}(r) = \exp(2\pi i H_j \cdot r), \quad j = 1, 2, \ldots, m,$$
$$C_{H_0}(r) = 1.$$

Let $\lambda_0, \lambda_1, \ldots, \lambda_m$ be a set of complex Lagrange multipliers. Then, we can form a Lagrangian function for the problem:

$$L(\rho, \lambda_0, \lambda_1, \ldots, \lambda_n) = S(\rho) + \sum_{j=0}^{n} \lambda_j \cdot \{F_{H_j}(\rho) - F_{H_j}^*\}. \tag{2.50}$$

Based on general optimization theory, a necessary condition for ρ to be a solution to problem (2.49) is that ρ must satisfy all of the constraints of the problem, and the partial derivative of the Lagrangian function with respect to ρ must also equal zero. Since the objective function of the problem is concave, the condition is also sufficient. It is easy to verify that the objective function for the problem is even strictly concave. Therefore, the solution must also be unique.

Note that

$$L'_{\rho}(\rho, \lambda_0, \lambda_1, \ldots, \lambda_n) = S'(\rho) + \sum_{j=0}^{n} \lambda_j \cdot F'_{H_j}(\rho).$$

Therefore, we have

$$S'(\rho)(\delta\rho) + \sum_{j=0}^{m} \lambda_j \cdot F'_{H_j}(\rho)(\delta\rho) = 0. \tag{2.51}$$

Since

$$S'(\rho)(\delta\rho) = -\int_V \left[\rho(r)\ln\left(\frac{\rho(r)}{q(r)}\right)\right]'(\delta\rho(r))\,dr$$

$$= -\int_V \left[\ln\left(\frac{\rho(r)}{q(r)}\right) + 1\right](\delta\rho(r))\,dr$$

and

$$F'_{H_j}(\rho)(\delta\rho) = -\int_V [\rho(r)C_{H_j}(r)]'(\delta\rho(r))\,dr$$

$$= -\int_V [C_{H_j}(r)](\delta\rho(r))\,dr,$$

we then have

$$-\int_V \left[\ln\left(\frac{\rho(r)}{q(r)}\right) + 1 - \sum_{j=0}^m \lambda_j \cdot C_{H_j}(r)\right](\delta\rho(r))\,dr = 0. \qquad (2.52)$$

Solve the equation to obtain

$$\rho(r) = q(r)\exp(\lambda_0 - 1)\exp\left[\sum_{j=1}^m \lambda_j \cdot C_{H_j}(r)\right]. \qquad (2.53)$$

Let $\lambda_0 = 1 - \log Z$ or, equivalently, $Z = \exp(1 - \lambda_0)$. Then,

$$\rho(r) = \frac{q(r)}{Z}\exp\left[\sum_{j=1}^m \lambda_j \cdot C_{H_j}(r)\right]. \qquad (2.54)$$

Since ρ satisfies all of the constraints and, in particular,

$$\int_V \rho(r)\,dr = \int_V \frac{q(r)}{Z}\exp\left[\sum_{j=1}^m \lambda_j \cdot C_{H_j}(r)\right]dr = 1,$$

we then obtain Z as a function of $\lambda_1, \lambda_2, \ldots, \lambda_m$:

$$Z(\lambda_1, \lambda_2, \ldots, \lambda_m) = \int_V q(r)\exp\left[\sum_{j=1}^m \lambda_j \cdot C_{H_j}(r)\right]dr. \qquad (2.55)$$

By applying all other constraints to ρ, we obtain

$$\int_V \frac{q(r)}{Z} \exp\left[\sum_{l=1}^{m} \lambda_l \cdot C_{H_l}(r)\right] C_{H_j}(r)\, dr = F^*_{H_j}, \quad j = 1, 2, \ldots, m.$$

$$(2.56)$$

Note that

$$\partial_j(\log Z) = \frac{1}{Z}\int_V q(r) \exp\left[\sum_{l=1}^{m} \lambda_l \cdot C_{H_l}(r)\right] C_{H_j}(r)\, dr,$$

where ∂_j represents $\partial_j/\partial\lambda_j$. Therefore, Eq. (2.56) can be written as

$$\partial_j(\log Z)(\lambda_1, \lambda_2, \ldots, \lambda_m) = F^*_{H_j}, \quad j = 1, 2, \ldots, m. \qquad (2.57)$$

This equation is called the entropy equation and can be used to determine all of the parameters $\lambda_1, \lambda_2, \ldots, \lambda_m$. Once these parameters are determined, ρ can be obtained with formula (2.53) or (2.54) and the entropy maximization problem is solved.

2.3.3. *Newton method*

We now discuss the solution of the entropy equation (2.57). Let g_j be a function such that

$$g_j(\lambda_1, \lambda_2, \ldots, \lambda_m) = \partial_j(\ln Z)(\lambda_1, \lambda_2, \ldots, \lambda_m) - F^*_{H_j}, \quad j = 1, 2, \ldots, m.$$

$$(2.58)$$

Then, we want to find $\lambda_1, \lambda_2, \ldots, \lambda_m$ so that

$$g_j(\lambda_1, \lambda_2, \ldots, \lambda_m) = 0, \quad j = 1, 2, \ldots, m. \qquad (2.59)$$

Let $\lambda = (\lambda_1, \lambda_2, \ldots, \lambda_m)^{\mathrm{T}}$, $g = (g_1, g_2, \ldots, g_m)^{\mathrm{T}}$, and $F^* = (F^*_{H_1}, F^*_{H_2}, \ldots, F^*_{H_m})^{\mathrm{T}}$. Then, the equations are equivalent to

$$g(\lambda) = \nabla(\log Z)(\lambda) - F^* = 0. \qquad (2.60)$$

This equation can be solved by using, for example, the Newton method. Thus, given any iterate λ, a new iterate λ^+ can be obtained with the formula

$$\lambda^+ = \lambda - [g'(\lambda)]^{-1} g(\lambda). \qquad (2.61)$$

Based on the theory for the Newton method, the formula can be used to generate a sequence of iterates that converges eventually to the solution of the equation if the initial iterate is sufficiently close to the solution. Also, the convergence rate is quadratic in the sense that the distance between the iterate λ and the solution λ^* reduces as a quadratic function in every step. On the other hand, in every step, a linear system of equations needs to be solved to obtain the Newton step

$$\Delta\lambda = -[g'(\lambda)]^{-1} g(\lambda), \tag{2.62}$$

which can be computationally costly because it generally requires $O(m^3)$ calculations. However, the coefficient matrix $g'(\lambda)$ actually has a special structure which can be exploited to obtain the solution of the equation more efficiently in $O(m \log m)$.

2.3.4. *Fast Fourier transform (FFT) and discrete convolution*

We now examine matrix $g'(\lambda)$ in Eq. (2.62). With the fact that

$$\rho(r) = \frac{q(r)}{Z} \exp\left[\sum_{j=0}^{m} \lambda_j \cdot C_{H_j}(r)\right],$$

it can be shown that

$$\partial_j(\log Z)(\lambda_1, \lambda_2, \ldots, \lambda_m) - F^*_{H_j} = \langle C_{H_j}\rangle - F^*_{H_j},$$

$$\partial_{jk}(\log Z)(\lambda_1, \lambda_2, \ldots, \lambda_m) = \langle C_{H_j} - C_{H_k}\rangle \overline{\langle C_{H_j} - C_{H_k}\rangle},$$

$$j, k = 1, 2, \ldots, m.$$

Putting these equations together, we have

$$g = F - F^*,$$

$$g' = K - FF^H, \tag{2.63}$$

where F^H is the complex conjugate of F,

$$F = (F_{H_1}, F_{H_2}, \ldots, F_{H_m})^{\mathrm{T}},$$

$$F^* = (F^*_{H_1}, F^*_{H_2}, \ldots, F^*_{H_m})^{\mathrm{T}},$$

and

$$K = \{K_{j,k}\} = \{F_{H_j - H_k}\}.$$

By using the Sherman–Morrison formula, we have

$$[g']^{-1} = (K - FF^H)^{-1} = K^{-1} + \frac{K^{-1}FF^H K}{1 - \sigma}, \qquad (2.64)$$

where $\sigma = F^H K^{-1} F$. The calculation of the Newton step then becomes

$$\Delta\lambda = -[g'(\lambda)]^{-1} g(\lambda) = V + \frac{F^H V}{1 - F^H U} U, \qquad (2.65)$$

where

$$V = K^{-1}(F - F^*),$$
$$U = K^{-1} F.$$

Note that

$$K = \{F_{H_j - H_k}\},$$
$$F_{H_j - H_k} = \int_V \rho(r) \exp\left[2\pi i (H_j - H_k) \cdot r\right] dr. \qquad (2.66)$$

Then, it is easy to verify that

$$K^{-1} = \{E_{H_j - H_k}\},$$
$$E_{H_j - H_k} = \int_V \rho^{-1}(r) \exp\left[2\pi i (H_j - H_k) \cdot r\right] dr. \qquad (2.67)$$

Therefore, the inverse of matrix K can actually be obtained directly from the Fourier transformation of the inverse of the density distribution function. By using fast Fourier transform (FFT), the calculation of K^{-1} can then be finished in $O(m \log m)$ floating-point operations.

For the calculation of the Newton step, we need not only K^{-1}, but also the product of K^{-1} and a vector, say X, which still requires $O(m^2)$ calculations in general. However, because the matrix K and its inverse are both circulant matrices, the product of such a matrix and a vector can also be reduced to $O(m \log m)$ floating-point operations.

Let $A = (A_1, A_2, \ldots, A_m)^T$ be a vector. Then, a circulant matrix $C(A)$ is a matrix with its first column equal to $(A_1, A_2, \ldots, A_m)^T$; the second, $(A_m, A_1, \ldots, A_{m-1})^{\mathrm{T}}$; the third, $(A_{m-1}, A_m, \ldots, A_{m-2})^{\mathrm{T}}$; and so on and so forth.

Let $A = (A_1, A_2, \ldots, A_m)^{\mathrm{T}}$ and $B = (B_1, B_2, \ldots, B_m)^{\mathrm{T}}$ be two vectors. A discrete convolution of A and B, denoted by $*$, is defined as

$$A * B = C(A)B, \tag{2.68}$$

where $C(A)$ is a circulant matrix generated by A. Now, let $a = (a_1, a_2, \ldots, a_m)^{\mathrm{T}}$ and $b = (b_1, b_2, \ldots, b_m)^{\mathrm{T}}$ be vectors such that $A = FT(a)$ and $B = FT(b)$, where FT represents Fourier transform and, in particular, discrete Fourier transform. Then, based on the discrete convolution theory,

$$A * B = FT(a) * FT(b) = FT(a \cdot b), \tag{2.69}$$

where $a \cdot b$ is the component-wise product of a and b. This simply implies that the discrete convolution of A and B can be calculated as a discrete Fourier transform for the component-wise product of a and b, which can actually be done in $O(m \log m)$ time. By applying this result to the product of $K^{-1}X$ for any vector $X = FT(d)$, we then have

$$K^{-1}X = E * X = FT(\rho^{-1}) * FT(d) = FT(\rho^{-1} \cdot b). \tag{2.70}$$

2.4. Indirect Methods

Direct methods have been applied successfully to small molecules, but not to macromolecules such as proteins. The reasons are not only that the mathematical problem becomes too difficult to solve for large molecules, but also that the diffraction data are not as accurate when there are too many atoms. Alternative approaches have been developed to utilize more structural information from various experimental techniques and have proven to be practical for solving protein structures.

2.4.1. *Patterson function*

Let ρ be an electron density distribution function, and

$$\rho(r) = \sum_{K \in Z^3} F_H \exp(-2\pi \mathrm{i} H \cdot r).$$

Then, a Patterson function P with respect to ρ is defined such that

$$P(u) = \int_V \rho(r)\rho(r + u)\, dr \tag{2.71}$$

for all u in R^3. It is easy to see that the Patterson function is essentially a convolution of ρ with itself at $-u$. Let $\rho(r + u)$ be expressed as

$$\rho(r + u) = \sum_{K \in Z^3} F_K \exp\left[-2\pi i K \cdot (r + u)\right].$$

Then,

$$\rho(r)\rho(r + u) = \sum_{H,K} F_H F_K \exp\{-2\pi i[(H + K) \cdot r + K \cdot u]\}.$$

Integrate the equation to obtain

$$\int_V \rho(r)\rho(r + u)\, dr$$

$$= \sum_{H,K} F_H F_K \exp(-2\pi i K \cdot u) \int_V \exp\left[-2\pi i(H + K) \cdot r\right] dr$$

$$= \sum_{H,K} F_H F_K \exp(-2\pi i K \cdot u) \int_V \cos\left[-2\pi(H + K) \cdot r\right] dr$$

$$= \sum_{H} F_H F_{-H} \exp(2\pi i H \cdot u).$$

Note that $F_H F_{-H} = |F_H|^2$, and we therefore have

$$P(u) = \int_V \rho(r)\rho(r + u)\, dr = \sum_H |F_H|^2 \cos(2\pi H \cdot u). \tag{2.72}$$

From the definition of the Patterson function, we see that $P(u)$ is the integral of the multiple of the electron densities at the beginning and ending positions of u. If the density at either side of u is zero, then the integral will be zero. On the other hand, if there are two atoms at the beginning and ending positions of u, then the densities will be high around the two positions, and the integral will not be zero and in fact will achieve a maximum. Because of this property, the Patterson function has a peak for any distance vector between two atoms, and the peaks can be used to detect the distance vectors for the atomic pairs in the molecule.

If the distance vectors for all of the atomic pairs are known, the positions of the atoms can in principle be determined, assuming that the vectors can be correctly assigned to the corresponding atomic pairs. The latter can be done if the molecule is small because a simple trial-and-error approach will reveal the best possible assignment. However, if the molecule is large, there are too many possible assignments to try, and it will be difficult to find the correct one. Furthermore, some distance vectors will not even be distinguishable.

2.4.2. *Isomorphous replacement*

The Patterson function is useful for the determination of the positions of a few specified atoms such as those required in the isomorphous replacement approach for structure determination. In this approach, a few heavy atoms, usually metal atoms, are introduced into the molecule so that they can occupy certain positions in the molecule while not greatly affecting the structure of the molecule when the molecule is crystallized.

Once the heavy atoms are introduced into the molecule, the electron densities around those atoms can be high, and the Patterson function will show large peaks at the distance vectors among them. Since there are only a few such atoms, the distance vectors can be assigned correctly to the atomic pairs after several trials and the positions of the atoms can then be determined.

After the positions of the heavy atoms are known, the positions of the remaining atoms and the phases of the structure factors can be derived using a conventional phase determination scheme, with the positions of the heavy atoms as starting positions.

2.4.3. *Molecular replacement*

An initial model for the molecule is usually required as a starting structure for the determination of the phases. Such a model can be obtained from some experimental methods such as the isomorphous replacement method or theoretical approaches such as homology modeling. Once a model becomes available, it can be used to generate initial phases and initial atomic positions. However, at first, the model structure must be rotated and translated so that it has the same position as the original crystal structure. Otherwise, the phases generated will not correspond to those from the crystal structure, because the values of the phases depend on the origin of the coordinate system selected for the crystal structure.

Let Q be a rotation matrix. For a given vector u, $u' = Qu$ is a vector obtained from rotating u by Q. A commonly used approach for rotating a given model is to compare the Patterson functions for the crystal structure and the model structure and see if they are correlated after the model is rotated. Consider

$$R = \int_V P(u) P\left(u'\right) du, \tag{2.73}$$

where $u' = Qu$. Then, we want to choose Q so that R can be maximized. Note that

$$P(u) = \sum_H |F_H|^2 \exp(2\pi i H \cdot u)$$

and

$$P(u') = \sum_K |F_K|^2 \exp(2\pi i K \cdot Qu).$$

Then,

$$R = \sum_{H,K} |F_H|^2 |F_K|^2 \int_V \exp[-2\pi i (H + Q^T K) \cdot u] du.$$

The integral in R is nonzero if $Q^T K = -H$ or $K = -QH$. So, the summation can be represented in a simpler form as

$$R = \sum_H |F_H|^2 |F_{-QH}|^2, \tag{2.74}$$

and an optimal Q can be found by calculating R repeatedly until it is maximized.

Appropriate translation of the model structure can also be obtained by comparing the Patterson functions of the crystal structure and the model structure. A commonly used approach is to first compute the cross-Patterson function

$$P_c(u, s) = \int_V \rho(x - s)\rho\left(Q^T[x + u - Qs - d]\right) dr$$

$$= \sum_H F_H F_{-QH} \exp\left[-2\pi i H \cdot (-s + Qs + d - u)\right].$$

Let $t = -s + Qs + d$, and redefine

$$P_c(u, t) = \sum_H F_H F_{-QH} \exp\left[-2\pi i H \cdot (t - u)\right].$$

Then, define a translation function

$$T = \int_V P_c(u, t) P(u)\, du, \tag{2.75}$$

and determine t so that T can be maximized. Note that the right-hand side of the formula is equal to

$$\sum_{H,K} F_H F_{-QH} |F_K|^2 \exp(-2\pi i H \cdot t) \int_V \exp\left[-2\pi i(K - H) \cdot u\right] du,$$

where each term vanishes unless $K = H$. Then, we have

$$T = \sum_H |F_H|^2 F_H F_{-QH} \exp(-2\pi i H \cdot t). \tag{2.76}$$

2.4.4. *Anomalous scattering*

Electrons around the atoms are not always free. For heavy atoms, the electrons in the inner shells are more bounded than those in the outer shells. The X-rays scattered by these atoms can therefore be altered. For example, the magnitudes of the structure factors for H and $-H$ can differ as a result of this so-called anomalous scattering.

In particular, the atomic scattering factor due to anomalous scattering will no longer be a real number. It may have a real part f_R and an imaginary part f_I, and

$$f = f_R + i f_I, \tag{2.77}$$

and the magnitude of structure factor F_H may not be equal to that of F_{-H}. However, the difference between the two factors can often be used for the determination of the positions of the heavy atoms.

With different wavelengths, the anomalous scattering produces different diffraction results. Let $F_{BA} = F_A + F_B$, where F_A is the anomalous contribution to F_{BA} and F_B is the nonanomalous contribution. Let F_H be the resulting structure factor. Then, in general,

$$|F_H|^2 = |F_{BA}|^2 + p|F_A|^2 + |F_{BA}||F_A|[q\cos(\phi_{BA}-\phi_A)+r\sin(\phi_{BA}-\phi_A)], \tag{2.78}$$

where p, q, r are functions of the wavelength λ, and $|F_{BA}|$ and $|F_B|$ as well as $(\phi_{BA} - \phi_A)$ are unknowns and are independent of the wavelengths. In each diffraction experiment, we can have two equations for each H, one for F_H and another for F_{-H}. By using several different wavelengths, we can have a multiple set of equations for F_H and F_{-H}. Since the number of equations is more than the number of unknowns, the unknowns can in principle be determined. In particular, the angles ϕ_{BA} can be determined, which is necessary for the determination of the phases for all of the corresponding structure factors F_H.

Selected Further Readings

The phase problem

Rhodes G, *Crystallography Made Crystal Clear*, Academic Press, 2000.

McPherson A, *Introduction to Macromolecular Crystallography*, Wiley, 2003.

Woolfson MM, *Introduction to X-ray Crystallography*, 2nd ed., Cambridge University Press, 2003.

Drenth J, *Principles of Protein X-ray Crystallography*, 3rd ed., Springer, 2006.

Least squares solutions

Sayre D, The squaring method: A new method for phase determination, *Acta Cryst* 5: 60–65, 1952.

Cochran W, A relation between the signs of structure factors, *Acta Cryst* 5: 65–67, 1952.

Woolfson MM, The statistical theory of sign relationships, *Acta Cryst* 7: 61–64, 1954.

Cochran W, Woolfson MM, The theory of sign relations between structure factors, *Acta Cryst* 8: 1–12, 1955.

Cochran W, Relations between the phases of structure factors, *Acta Cryst* 8: 473–478, 1955.

Germain G, Woolfson MM, On the application of phase relationships to complex structures, *Acta Cryst* **B24**: 91–96, 1968.

German G, On the application of phase relationships to complex structures II. Getting a good start, *Acta Crsyt* **B28**: 274–285, 1970.

Hauptman H, Karle J, Crystal-structure determination by means of a statistical distribution of interatomic vectors, *Acta Cryst* 5: 48–59, 1952.

Hauptman H, Karle J, *Solution of the Phase Problem I. The Centrosymmetric Crystal*, American Crystallography Association Monograph, Vol. 3, Polycrystal Book Service, 1953.

Karle J, Karle IL, The symbolic addition procedure for phase determination for centrosymmetric and noncentrosymmetric crystals, *Acta Cryst* **21**: 849–859, 1966.

Ladd MFC, Palmer RA, *Theory and Practice of Direct Methods in Crystallography*, Plenum Press, 1980.

Sayre D, *Computational Crystallography*, Oxford University Press, 1982.

DeTitta GT, Weeks CM, Thuman P, Miller R, Hauptman H, Structure solution by minimal-function phase refinement and Fourier filtering. I. Theoretical basis, *Acta Cryst* **A50**: 203–210, 1994.

DeTitta GT, Weeks CM, Thuman P, Miller R, Hauptman H, Structure solution by minimal-function phase refinement and Fourier filtering. II. Implementation and applications, *Acta Cryst* **A50**: 210–220, 1994.

Hauptman H, A minimal principle in the phase problem of X-ray crystallography, in *Global Minimization of Nonconvex Energy Functions: Molecular Conformation and Protein Folding*, Pardalos PM, Shalloway D, Xue G (eds.), American Mathematical Society, pp. 89–96, 1996.

Entropy maximization

Klug A, Joint probability distributions of structure factors and the phase problem, *Acta Cryst* **11**: 515–543, 1958.

Piro OE, Information theory and the "phase problem" in crystallography, *Acta Cryst* **A39**: 61–68, 1983.

Bricogne G, Maximum entropy and the foundations of direct methods, *Acta Cryst* **A40**: 410–445, 1984.

Bricogne G, A Bayesian statistical theory of the phase problem. I. A multi-channel maximum-entropy formalism for constructing generalized joint probability distributions of structure factors, *Acta Cryst* **A44**: 517–545, 1988.

Bricogne G, Direct phase determination by entropy maximization and likelihood ranking: Status report and perspectives, *Acta Cryst* **D49**: 37–60, 1993.

Bricogne G, A Bayesian statistical viewpoint on structure determination: Basic concepts and examples, *Methods Enzymol* **276**: 361–423, 1997.

Bricogne G, Gilmore CJ, A multisolution method of phase determination by combined maximization of entropy and likelihood. I. Theory, algorithms and strategy, *Acta Cryst* **A46**: 284–297, 1990.

Gilmore CJ, Bricogne G, Bannister C, A multisolution method of phase determination by combined maximization of entropy and likelihood. II. Applications to small molecules, *Acta Cryst* **A46**: 297–308, 1990.

Wu Z, Phillips G, Tapia R, Zhang Y, A fast Newton's method for entropy maximization in statistical phase estimation, *Acta Cryst* **A57**: 681–685, 2001.

Wu Z, Phillips G, Tapia R, Zhang Y, A fast Newton's algorithm for entropy maximization in phase determination, *SIAM Rev* **43**: 623–642, 2001.

Indirect methods

Blow DM, Matthews BW, Parameter refinement in the multiple isomorphous-replacement method, *Acta Cryst* **A29**: 56–62, 1973.

Watenpaugh KD, Overview of phasing by isomorphous replacement, *Methods Enzymol* **115**: 3–15, 1985.

Rossmann MG, *The Molecular Replacement Method*, Gordon & Breach, 1972.

Dodson EJ, Gover S, Wolfe W, *Molecular Replacement*, SERC Daresbury Laboratory, Warrington, UK, 1992.

Hendrickson WA, Smith JL, Sheriff S, Direct phase determination based on anomalous scattering, *Methods Enzymol* **115**: 41–55, 1985.

Hendrickson WA, Ogata CM, Phase determination from multiwavelength anomalous diffraction measurements, *Methods Enzymol* **276**: 494–523, 1997.

Chapter 3

NMR Structure Determination

Nuclear magnetic resonance (NMR) spectroscopy is another important experimental technique for protein structure determination. Of about 30 000 protein structures deposited in the Protein Data Bank (PDB) so far, 15% were determined by NMR. The advantage of using NMR for structure determination is that it does not require crystallizing the protein and the structure can be determined in solution, which is critical for proper folding.

NMR is used to detect some conformational properties of proteins such as the distances between certain pairs of atoms, upon which the structures can be modeled. The solution to the problem of determining the coordinates of the atoms in the protein given various NMR-detected distance constraints is essential to NMR modeling. An abstract form of the problem is called the distance geometry problem in mathematics, the graph embedding problem in computer science, and the multidimensional scaling problem in statistics.

3.1. Nuclear Magnetic Resonance

A nucleus, with electrons spinning around it, can generate a small magnetic field. This happens especially for nuclear isotopes like ^{1}H, ^{13}C, and ^{15}N. When such a nucleus is placed in an external magnetic field, it will act as a small magnet and try to align itself with the external field. There can be two relatively stable states for the nucleus: when it is in the same direction as the external field or opposite to it. The former is called the upward state, and the latter the downward state. Of the two, the downward state has the higher energy.

When the nucleus changes its state, it absorbs or releases energy known as absorption energy. Different nuclear isotopes have different absorption energies. The nuclei of the same isotope may also have different absorption energies when they are in different chemical or physical environments. The absorption energies of nuclear isotopes can be detected with NMR techniques and used to derive various physical or geometrical properties of molecules.

3.1.1. *Nuclear magnetic fields*

A nuclear isotope such as ^{1}H has an intrinsic source of angular momentum called nuclear spin angular moment. This angular momentum generates a small magnetic moment called the nuclear spin magnetic moment. Let M_0 be a vector representing the external magnetic field and M the magnetic moment of the nucleus. The interaction between the nuclear magnetic moment and the external magnetic field has the lowest energy if M is in the same direction as M_0, and the highest energy if M is in the opposite direction of M_0. If M and M_0 are not parallel and have an angle θ between them, the interaction between M and M_0 generates an angular moment with M rotating around M_0, sweeping out a cone with constant angle θ. This motion is called precession, and M is said to process about M_0 (Fig. 3.1).

Let B_0 be the magnetic strength of M_0. Let ω_0 be the angular frequency of M precessing about M_0. Then, ω_0 is proportional to B_0 and

$$\omega_0 = \gamma B_0, \tag{3.1}$$

where γ is called the gyromagnetic ratio. It turns out that the difference between the two energy levels of M (the highest and the lowest) or, in other words, the absorption energy of M is proportional to the precession frequency ω_0 of M about M_0. Therefore, in order to find the absorption energies of nuclear isotopes, we can just focus on their precession frequencies.

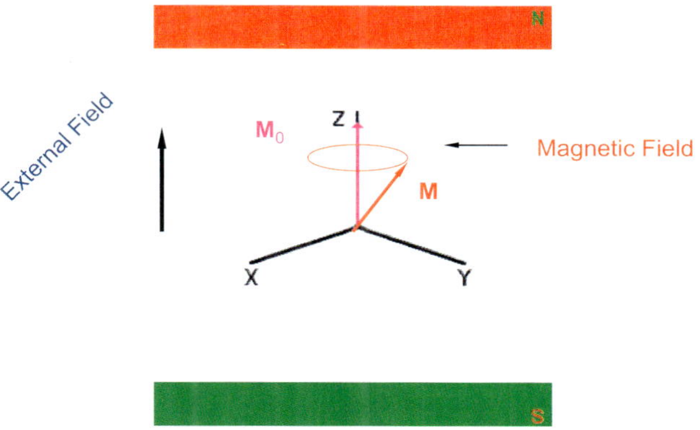

Fig. 3.1. Nuclear spin angular momentum. The interaction between the nuclear magnetic moment and the external magnetic field has the lowest energy if M is in the same direction of M_0, and the highest energy if M is in the opposite direction of M_0.

Let M_0 be placed along the z-axis in a three-dimensional (3D) coordinate system. If we apply another small magnetic field M_1 to the nucleus along a direction perpendicular to M_0, say along the x-axis, the nuclear magnetic moment M will tilt slightly toward M_1 and try to precess about M_1 as well. It turns out that if we can rotate the x–y plane around the z-axis with the same speed or, more accurately, the same frequency as the precession frequency of M about M_0, the magnetic moment M will act as if it only interacts with M_1 and rotate around the x-axis.

In fact, the rotation of the x–y plane around the z-axis can be achieved indirectly by applying a periodically changing magnetic field M_1 along the x-axis (Fig. 3.2). Such a field can be produced by sending an electrical pulse in a coil wired around the x-axis. Let ω_0 be the precession frequency of M about M_0, and ω_1 the frequency of the pulse and hence the frequency of the magnetic field M_1. If ω_1 is equal to ω_0 (in resonance), M_1 changes with the same frequency as the precession frequency of M about M_0, which is equivalent to rotating the x–y plane around the z-axis with the same frequency as the precession frequency of M about M_0. As a result, if M is in the same direction along the z-axis, it will precess about M_1 and flip around the x-axis

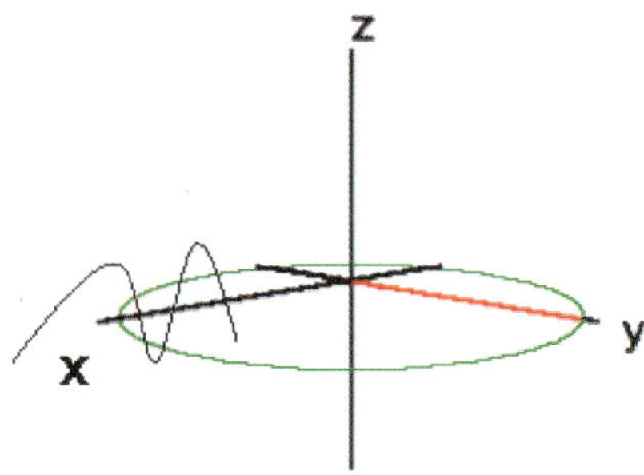

Fig. 3.2. Electrical and magnetic signals. The precession of nuclear magnetic field around the external magnetic field produces an electrical signal, which can then be transformed to an NMR spectrum in frequency domain.

in the y–z plane. Note that the precession frequency is proportional to the magnetic strength of M_1, and is therefore equal to γB_1.

A particular way of making a resonance is that if we let the pulse to sustain for a time t, M will rotate at an angle $\beta = \gamma B_1 t$ from z to $-z$ counter clockwise. In particular, M will flip from z to $-y$ if $t = \pi/(2\gamma B_1)$, and from z to $-z$ if $t = \pi/(\gamma B_1)$. The pulses used are generally called the $90°$ and $180°$ pulses, respectively, or the on-resonance pulses.

In practice, due to the different chemical or physical conditions they have, different nuclei tend to have slightly different precession frequencies. Therefore, ω_1 may not always be equal to ω_0. However, as long as ω_1 is close to ω_0, M will still make similar flips around the x-axis as described above. The pulse making such flips is called the hard-resonance pulse.

Let $\Omega_0 = \omega_0 - \omega_1 \neq 0$. Effectively, there will still be a magnetic moment along the z-axis but with a reduced precession frequency Ω_0. Then, after M makes a flip, say from z to $-y$, it will continue to precess about M_0 with the frequency Ω_0 in the x–y plane. After a time t, M will move to a position with

$$M_x = |M| \sin \Omega_0 t,$$
$$M_y = -|M| \cos \Omega_0 t. \tag{3.2}$$

Now, when M rotates around the z-axis in the x–y plane, it cuts through the coil around the x-axis and induces an electrical current in the coil. The latter should have the same frequency as Ω_0.

Suppose that there are m nuclei of the same type in a given molecule, with precession frequencies $\omega_{01}, \omega_{02}, \ldots, \omega_{0m}$. Let $\Omega_{01} = \omega_{01} - \omega_1$, $\Omega_{02} = \omega_{02} - \omega_1, \ldots, \Omega_{0m} = \omega_{0m} - \omega_1$, where ω_1 is the oscillating frequency of M_1. Then, all together, the nuclei make a magnetic moment M such that at time t,

$$M_x = |M_{01}| \sin \Omega_{01} t + |M_{02}| \sin \Omega_{02} t + \cdots + |M_{0m}| \sin \Omega_{0m} t \quad (3.3)$$

and

$$M_y = -|M_{01}| \cos \Omega_{01} t - |M_{02}| \cos \Omega_{02} t - \cdots - |M_{0m}| \cos \Omega_{0m} t, \quad (3.4)$$

where $M_{01}, M_{02}, \ldots, M_{0m}$ are the magnetic moments generated by nuclei $1, 2, \ldots, m$. The magnetic moment M induces an electric current in the coil, with frequencies $\Omega_{01}, \Omega_{02}, \ldots, \Omega_{0m}$.

3.1.2. *NMR spectra*

An NMR spectrum shows the distribution of the nuclear magnetic strength over a range of frequencies for a given molecule. Such a distribution can be extracted from the electrical signal generated by the nuclei in the molecule in an NMR experiment, as described in the previous section.

Let f be a function of time t that describes the intensity of the electrical signal. Let F be a function of frequency Ω that gives the frequency distribution of the signal. Then, F can be obtained by applying a Fourier transform on f in real space; and formally, for any frequency Ω,

$$F(\Omega) = \int_{-\infty}^{+\infty} f(t) \cos(\Omega t) \, dt, \quad (3.5)$$

where Ω is in fact the difference between the frequency and a reference frequency.

Let ω_0 be the precession frequency of a nucleus in the molecule and ω_1 the reference frequency, where ω_0 and ω_1 have the same meaning as described in the previous section. Then, the Fourier transform F of f achieves a maximum at the corresponding frequency difference $\Omega_0 = \omega_0 - \omega_1$. The maxima of F show the frequencies at which NMR occurs in the molecule, and therefore form an NMR spectrum for the molecule.

Each intensity peak in the NMR spectrum corresponds to a nucleus in the molecule precessing about the external field with the frequency of the peak. The precession frequency is proportional to the energy difference between the upward and downward states of the nucleus. Therefore, the peaks in the spectrum can also be viewed as the indicators for energy transitions of the molecule at different energy levels, with each corresponding to the transition between two energy states of a particular nucleus inside the molecule. In any case, because of the relationship between the frequency distribution in the NMR spectrum and the presence of certain nuclei in the molecule, the NMR spectral data can be used to derive various physical or geometrical properties of the nuclei in the molecule, such as the types of nuclei, the chemical groups they belong to, their locations, the distances among them, etc. All of these properties are important for determining the structure of the molecule.

As we have discussed in the previous section, the precession frequency of a nucleus is proportional to the strength of the magnetic field. Therefore, if the strength of the magnetic field is changed, the NMR spectrum will also be changed. In order to remove the dependency on the frequency, in practice, the NMR spectrum is often defined as a relative frequency called chemical shift (ppm) instead of absolute frequency. The relative frequency is defined as the difference of the frequency from a reference frequency divided by the reference frequency, which is frequency-independent. Naturally, the frequency of the oscillating magnetic field applied in a horizontal direction can be used as a reference frequency.

Different types of nuclear isotopes have different precession frequencies. Even the same type of nuclei may have different precession frequencies if they are in different chemical groups, but the differences are not as great as for different types of nuclear isotopes. In practice, experiments are often conducted to obtain the NMR spectrum for one type of nuclear isotope at a time, for example, for ^{1}H, and then ^{13}C, and then ^{15}N. For each particular type, the spectrum can be observed within a certain range of frequencies. For example, for ^{1}H, the frequencies may range from 1 ppm to 20 ppm; while for ^{13}C, the range may be from 20 ppm to 200 ppm.

NMR intensities are rather weak and subject to experimental errors. The experiments are therefore often repeated many times to increase the signal–noise ratio and hence the accuracy of the spectra. On the other hand, since the results from NMR denote the average behaviors of the system over time, they are often considered to be more representative of the true dynamic nature of the system than other experimental approaches.

The shape and position of the peak for a nucleus in the NMR spectrum may also be affected when there is another nucleus nearby, especially when they are connected by chemical bonds. The influence also differs depending on the state of the neighboring nucleus. If the nearby nucleus is in its upward state, the frequency peak tends to shift slightly to the left; alternatively, if the nearby nucleus is in its downward state, the peak tends to shift to the right. As a result, the original peak may be split into two separate peaks, and peak splitting can in fact indicate the existence of a bonded and spatially close nuclear pair. The interaction between such a pair is called *J*-coupling.

3.1.3. *COSY experiment*

A COSY experiment, or NMR correlation spectroscopy, is designed to obtain a two-dimensional (2D) rather than one-dimensional (1D) NMR spectrum so that the NMR intensities can be observed for two varying frequencies. Many different 2D NMR experiments can be designed, but the most important ones for structure determination are COSY and NOE experiments. The latter will be discussed in the next section.

A 2D NMR spectrum, defined as a function $F(\omega_1,\omega_2)$ of frequency variables ω_1 and ω_2, can be obtained by performing a Fourier transform twice on a function $S(t_1, t_2)$ of two time variables t_1 and t_2 that represents the electrical signal recorded from the NMR experiment

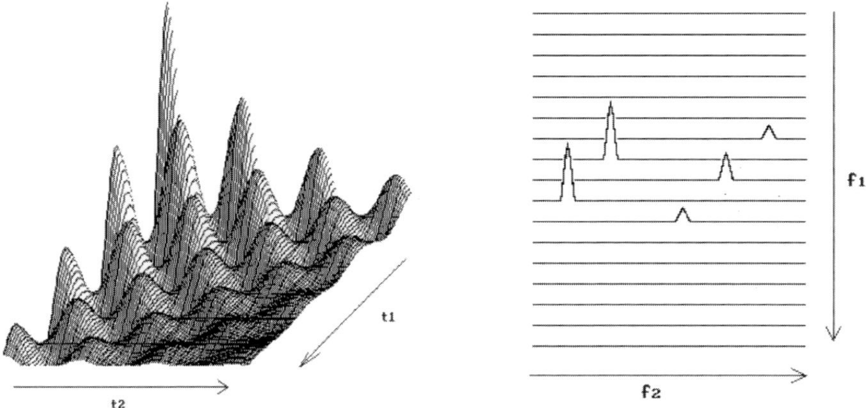

Fig. 3.3. Two-dimensional spectrum. A 2D NMR spectrum can be obtained by performing a Fourier transform twice on a function of two time variables t_1 and t_2 that represents the electrical signal recorded from the NMR experiment.

(Fig. 3.3). Mathematically,

$$F(\omega_1, \omega_2) = \int_{-\infty}^{+\infty} \int_{-\infty}^{+\infty} S(t_1, t_2) \exp[-i(\omega_1 t_1 + \omega_2 t_2)] dt_2 \, dt_1. \quad (3.6)$$

Usually, t_2 is the time when the NMR signal is recorded, and t_1 is a special time period before t_2. The function $S(t_1,t_2)$ is obtained by recording the NMR signal in t_2 while varying t_1. So, for each fixed t_1, a signal is recorded for t_2. This signal can be transformed to obtain $F_S(t_1,\omega_2)$:

$$F_S(t_1, \omega_2) = \int_{-\infty}^{+\infty} S(t_1, t_2) \exp[-i(\omega_2 t_2)] \, dt_2. \quad (3.7)$$

For each fixed ω_2, $F_S(t_1,\omega_2)$ varies with t_1 and can also be transformed to obtain $F(\omega_1,\omega_2)$:

$$F(\omega_1, \omega_2) = \int_{-\infty}^{+\infty} F_s(t_1, \omega_2) \exp[-i(\omega_1 t_1)] \, dt_1. \quad (3.8)$$

For example, a pulse sequence for detecting J-coupling has the following steps: (1) a 90° pulse; (2) a $t_1/2$ delay time; (3) a 180° pulse;

(4) another $t_1/2$ delay time; and (5) an acquisition time t_2. The NMR signal is recorded in the acquisition time t_2, but changes as the delay time t_1 varies. More accurately, the NMR strength produced by this experiment is equal to

$$|M| \cos(Jt_1/2) \cos(\Omega t_2), \tag{3.9}$$

which translates into an electrical signal $S(t_1, t_2)$ of variables t_1 and t_2, where J is called the J-coupling constant. From the above formula, in the 2D spectrum, at $\omega_1 = \pm J/2$ and $\omega_2 = \Omega$, there will be a peak, indicating that a nucleus at frequency Ω has a J-coupling with coupling constant J.

The COSY experiment is an effective way of finding the correlations between two NMR spectra, including the J-couplings among the nuclei. A simple COSY experiment for detecting ^1H–^1H coupling can be designed as follows: (1) a $90°$ pulse; (2) a t_1 delay time; (3) a $90°$ pulse; and (4) an acquisition time t_2. With such a pulse sequence, an NMR signal $S(t_1, t_2)$ can be obtained by varying t_1 and t_2. For a pair of coupled nuclei, the signal strength is typically given by the formula

$$|M| \cos(Jt_1/2) \cos(\Omega_\text{H} t_1) \cos(Jt_2/2) \cos(\Omega_\text{H} t_2), \tag{3.10}$$

where Ω_H is the precession frequency of ^1H. From this formula, it is clear that in the NMR spectra, there will be peaks at $\omega_1 = \Omega_\text{H} \pm J$ and $\omega_2 = \Omega_\text{H} \pm J$, showing that a J-coupling must have occurred between the corresponding ^1H nuclei.

3.1.4. *Nuclear Overhauser effect (NOE)*

Two nuclei interact with each other when they are close in space, whether or not they are chemically bonded. If they are bonded, the interaction is called spin–spin coupling or J-coupling, which is what we have discussed in the previous section. The other type of interaction is called dipole coupling, when there is not necessarily a bond connecting the nuclei.

The principle behind nuclear dipole coupling is that when two nuclei are close in space, the resonance frequency of one nucleus will

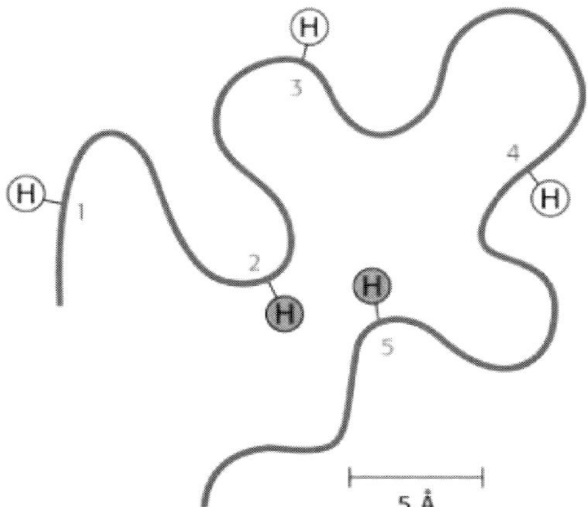

Fig. 3.4. Nuclear Overhauser effect. When two nuclei are close in space, the resonance frequency of one nucleus will be affected because of the presence of the other nucleus. The change in frequency is called the nuclear Overhauser effect (NOE) and can be measured through an NMR experiment.

be affected due to the presence of the other nucleus. The change in frequency is called the nuclear Overhauser effect (NOE) and can be measured through an NMR experiment. However, the NOE signal is generally weak and hard to detect when the two nuclei are not so close (Fig. 3.4).

Suppose that there are two nuclei I and S. The NOE on one of the nuclei I can be estimated using the formula

$$\eta = \frac{\gamma_S}{\gamma_I} \frac{W_2 - W_0}{2W_{1I} + W_2 + W_0}, \tag{3.11}$$

where η is called the NOE factor; γ_I and γ_S are the gyromagnetic ratios of I and S, respectively; and W_0, W_{1I}, and W_2 are constants

$$W_0 = \frac{1}{10}\kappa^2 \frac{\tau_c}{1 + (\omega_I - \omega_S)^2 \tau_c^2}, \tag{3.12}$$

$$W_{1I} = \frac{2}{20}\kappa^2 \frac{\tau_c}{1 + \omega_I^2 \tau_c^2}, \tag{3.13}$$

$$W_2 = \frac{3}{5}\kappa^2 \frac{\tau_c}{1 + (\omega_I + \omega_S)^2 \tau_c^2}, \tag{3.14}$$

where τ_c is a constant; ω_I and ω_S are resonance frequencies of I and S, respectively; and κ is inversely proportional to the third power of the distance $r_{I,S}$ between I and S. It follows that the accuracy of the NOE factor η is inversely proportional to the sixth power of the distance $r_{I,S}$, and it may only be detected for nearby nuclei, usually within <5 Å distance.

The pairs of nuclei that have NOE between them can be detected from a 2D NMR experiment called NOESY, referring to NOE spectroscopy. A typical pulse sequence for obtaining NOESY is as follows: (1) a $90°$ pulse; (2) a t_1 delay time; (3) a $90°$ pulse; (4) a τ mixing time; (5) a $90°$ pulse; and (6) an acquisition time t_2. By using such a pulse sequence, the NMR signal $S(t_1,t_2)$ can be obtained as a function of t_1 and t_2:

$$S(t_1,t_2) = c|M_I| \cos \Omega_I t_1 \exp(i\Omega_S t_2), \qquad (3.15)$$

where c is a constant; and Ω_I and Ω_S are the resonance frequencies of I and S, respectively. From this formula, we see that the 2D spectrum of the signal will have a peak at $\omega_1 = \Omega_I$ and $\omega_2 = \Omega_S$, showing that an NOE must have occurred between I and S.

3.2. Distance Geometry

The determination of the coordinates of the atoms in the molecule once a set of distance data becomes available from the NMR experiments is essential to the NMR techniques for structure determination. The nature of the data is that the distances are short with lengths usually less than or equal to 5 Å (due to the weak NOE signals that can be detected only for nearby nuclei), and they are available only for pairs of hydrogen atoms a short distance apart. Such a set of data contains important structural information, but it is not sufficient for the complete determination of the structure. Fortunately, with general knowledge of the bond lengths and bond angles, an additional set of distance data can be obtained. By combining the two sets of data, the coordinates of the atoms can then be determined, upon which a model for the molecule can be constructed.

3.2.1. *The fundamental problem*

The problem of determining the coordinates of the atoms in a molecule given the distances between certain pairs of atoms can be studied in a general mathematical form, where the atoms can be placed in any metric space with the distances defined in terms of a general metric associated with the space (Fig. 3.5). The problem may or may not have a solution, depending on the given distance data and the space where the solution is to be found. When it exists, the solution may still be nonunique. These properties carry great theoretical and practical importance, but have not been well understood. We will consider the problem only in Euclidean space, in particular the 3D Euclidean space, where the problem for molecular modeling is defined.

Suppose that we have a molecule of n atoms. Let x_j, $j = 1, 2, \ldots, n$ be a set of 3D vectors, where x_j is the coordinate vector for atom j. Let S be a set of index pairs (i, j), with each corresponding to a given distance $d_{i,j}$ between atoms i and j. Then, the problem of determining the coordinates of the atoms in the molecule with a set of given distances can be stated formally as follows:

$$||x_i - x_j|| = d_{i,j}, \quad (i, j) \in S, \tag{3.16}$$

where $|| \cdot ||$ represents the Euclidean norm. Therefore, for any $x_j = (x_{j,1}, x_{j,2}, x_{j,3})^{\mathrm{T}}$,

$$||x_j|| = \sqrt{x_{j,1}^2 + x_{j,2}^2 + x_{j,3}^2}. \tag{3.17}$$

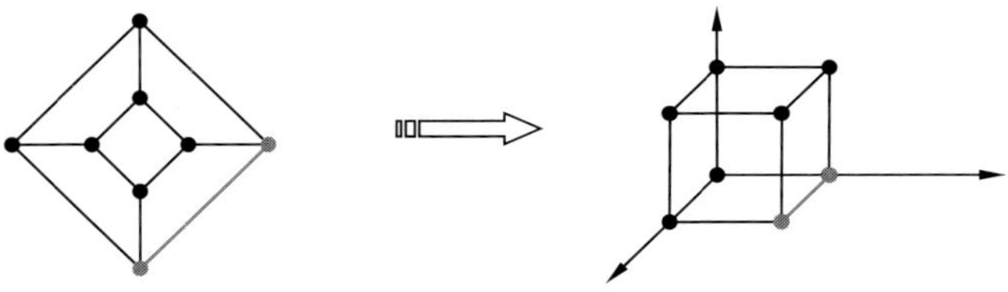

Fig. 3.5. Distance geometry problem. Given n atoms $a_1, a_2, \ldots, a_n$ and a set of distances $d_{i,j}$ between a_i and a_j, find positions $x_1, x_2, \ldots, x_n$ for $a_1, a_2, \ldots, a_n$ such that $||x_i - x_j|| = d_{i,j}$.

The set S may not necessarily contain all possible i,j pairs. We say that the problem has a sparse set of distances, or sparse distance data, if S has only a subset of all i,j pairs; otherwise, we say that it has a complete set of distances or dense distance data. The distances may not be provided as exact values and, in many cases, may be given in estimated ranges. When the exact distances are provided, we say that the problem has exact distances; otherwise, it has inexact distances or distance ranges or bounds. In the latter case, the solution is generally not unique, and there may in fact exist a set of solutions that may all be of interest in practice.

3.2.2. *Exact distances*

Assume that a complete set of exact distances is given. The distance geometry problem can then be solved in polynomial-time. A classical solution to the problem is based on the singular value decomposition (SVD) of a transformed distance matrix.

Define the coordinate and distance matrices for the molecule by

$$X = \{x_{i,j} : i = 1, 2, \ldots, n, \ j = 1, 2, 3\}, \tag{3.18}$$

$$D = \{d_{i,j} : i, j = 1, 2, \ldots, n\}, \tag{3.19}$$

respectively. Then, the distance geometry problem for a complete set of exact distances is essentially to find X given D.

Assume that a solution X exists for a given D. Then, $||x_i - x_j|| = d_{i,j}$ for all $i, j = 1, 2, \ldots, n$, and

$$||x_i||^2 - 2x_i^T x_j + ||x_j||^2 = d_{i,j}^2, \quad i, j = 1, 2, \ldots, n.$$

Since the structure is invariant under any translation or rotation, we set a reference system so that the origin is located at the last atom or, in other words, $x_n = (0, 0, 0)^T$. It follows that

$$d_{i,n}^2 - 2x_i^T x_j + d_{j,n}^2 = d_{i,j}^2, \quad i, j = 1, 2, \ldots, n - 1.$$

Redefine the coordinate and distance matrices X and D so that

$$X = \{x_{i,j} : i = 1, 2, \ldots, n - 1, j = 1, 2, 3\}, \qquad (3.20)$$

$$D = \{(d_{i,n}^2 - d_{i,j}^2 + d_{j,n}^2)/2 : i, j = 1, 2, \ldots, n - 1\}. \qquad (3.21)$$

Then, $XX^{\mathrm{T}} = D$ and D must be of maximum rank 3. We call this matrix D the transformed distance matrix. In general, if D is transformed from a distance matrix in k-dimensional Euclidean space, it must be of maximum rank k.

The equation $XX^{\mathrm{T}} = D$ can be solved by using SVD. Let $D = U\Sigma U^{\mathrm{T}}$ be the SVD of D, where U is an orthogonal matrix and Σ is a diagonal matrix with the singular values of D along the diagonal. If D is a matrix of rank less than or equal to 3, the decomposition can be obtained with U being $(n - 1) \times 3$ and Σ being 3×3. Then, $X = U\Sigma^{1/2}$ solves the equation $XX^{\mathrm{T}} = D$. Otherwise, if the rank of D is greater than 3, there will be no solution to the equation. Here, the SVD requires only $O(n^2)$ floating-point operations, and therefore the distance geometry problem is said to be polynomial-time solvable.

3.2.3. *Sparse distances*

We still consider the exact distances, but assume that they are available only for a subset of all possible pairs of atoms. Then, the distance geometry problem becomes difficult to solve. In computer science terminology, the problem can be proven to be NP-hard. We use a 1D version of the problem to demonstrate this property.

Suppose that we have a molecule of n atoms. Let x_j, $j = 1, 2, \ldots, n$ be a set of positions on a real line, where x_j is the position for atom j. Let S be a set of index pairs (i,j), with each corresponding to a given distance $d_{i,j}$ between atoms i and j. Then, a 1D distance geometry problem is to determine the positions x_j of the atoms in the molecule on a real line so that

$$|x_i - x_j| = d_{i,j}, \quad (i, j) \in S. \qquad (3.22)$$

We show that an integer set partition problem can be reduced to a 1D distance geometry problem.

Let s_j, $j = 1, 2, \ldots, n$ be a set of positive integers. An integer set partition problem is to find two subsets of the integers so that the sums of the integers in the subsets are equal, i.e.

$$\sum_{j \in S_1} s_j = \sum_{j \in S_2} s_j, \tag{3.23}$$

where S_1 and S_2 are the index sets of the integers in the first and second subsets, respectively.

Given an above integer set partition problem, we can always construct a 1D distance geometry problem for $n+1$ atoms with a set of distances as given below:

$$d_{j,j+1} = s_j, \quad 1 \le j \le n, \quad d_{1,n+1} = 0. \tag{3.24}$$

If the distance geometry problem has a solution, then the constraint $d_{1,n+1} = 0$ implies that $x_{n+1} = x_1$. Thus,

$$\sum_{j=1}^{n} (x_{j+1} - x_j) = x_{n+1} - x_1 = 0. \tag{3.25}$$

Since $|x_{j+1} - x_j| = s_j$, $x_{j+1} - x_j = +s_j$ or $-s_j$. Let $S_1 = \{j : x_{j+1} - x_j = +s_j\}$ and $S_2 = \{j : x_{j+1} - x_j = -s_j\}$. Then,

$$\sum_{j \in S_1} (x_{j+1} - x_j) + \sum_{j=S_2} (x_{j+1} - x_j) = \sum_{j \in S_1} s_j - \sum_{j=S_2} s_j = 0$$

and

$$\sum_{j \in S_1} s_j = \sum_{j \in S_2} s_j,$$

showing that the two subsets of integers with indices in S_1 and S_2 solve the original integer set partition problem.

The results above just show that the solution to a set partition problem can always be obtained by solving an equivalent 1D distance geometry problem. However, as we have already learned in computational theory, the integer set partition problem is an NP-hard problem, and therefore the distance geometry problem cannot be solved in polynomial time; otherwise, the integer set partition problem would be polynomial time solvable, contradicting the fact that the latter is in fact NP-hard.

3.2.4. *Distance bounds*

In NMR, the distances are often provided with some estimated bounds. The distance geometry problem becomes finding coordinates x_j of the atoms $j = 1, 2, \ldots, n$ so that the distances $d_{i,j}$ between atoms i and j are within their estimated lower and upper bounds, $l_{i,j}$ and $u_{i,j}$, respectively, for (i,j) in a subset S of all i, j pairs. That is,

$$l_{i,j} \leq ||x_i - x_j|| \leq u_{i,j}, \quad (i, j) \in S. \tag{3.26}$$

Let $d_{i,j} = (l_{i,j} + u_{i,j})/2$ and $\varepsilon_{i,j} = (u_{i,j} - l_{i,j})/2$. We can rewrite the above problem as

$$|\,||x_i - x_j|| - d_{i,j}| \leq \varepsilon_{i,j}, \quad (i, j) \in S. \tag{3.27}$$

Then, the problem can be viewed as finding an approximate solution to the distance geometry problem for a set of exact distances $d_{i,j}$ with each distance $d_{i,j}$ to be allowed for an error $\varepsilon_{i,j}$. We call such a solution an ε-approximation.

If large errors are allowed, an approximate solution is certainly easier to obtain than an exact solution. However, if only small errors are allowed, the problem of finding an approximate solution can be as hard as finding an exact solution. To see this, we first consider the problem of finding an approximate solution to the integer set partition problem.

Let $s_j, j = 1, 2, \ldots, n$ be a set of positive integers. An approximate solution to the integer set partition problem for the given set of integers is to find two subsets of real numbers $t_j, j = 1, 2, \ldots, n$ such that

$$\left| \sum_{j \in S_1} t_j - \sum_{j \in S_2} t_j \right| \leq \frac{1}{2}, \quad |t_j - s_j| \leq \varepsilon_j, \quad j = 1, 2, \ldots, n, \tag{3.28}$$

where S_1 and S_2 are the index sets of the numbers in the first and second subsets, respectively, and ε_j are the differences allowed for the real numbers from the corresponding integers.

Suppose that S_1 and S_2 give a partition for the numbers t_j, $j = 1, 2, \ldots, n$ and hence an approximate solution to the corresponding integer set partition problem. Then,

$$
\left| \sum_{j \in S_1} s_j - \sum_{j \in S_2} s_j \right| \leq \left| \sum_{j \in S_1} (s_j - t_j) - \sum_{j \in S_2} (s_j - t_j) \right| + \left| \sum_{j \in S_1} t_j \right.
$$

$$
\left. - \sum_{j \in S_2} t_j \right| \leq \sum_{j \in S_1} |s_j - t_j| + \sum_{j \in S_2} |s_j - t_j| + 1/2.
$$

Assume that $\varepsilon_j < 1/2n$ for all $j = 1, 2, \ldots, n$. The above result implies that

$$
\left| \sum_{j \in S_1} s_j - \sum_{j \in S_2} s_j \right| < \frac{1}{2} + \frac{1}{2} = 1.
$$

Note that the sums in the above inequality are over integers. Therefore, if the difference between the two sums is less than 1, the two sums must be equal. It follows that S_1 and S_2 give a partition for the integers s_j, $j = 1, 2, \ldots, n$ and hence an exact solution to the integer set partition problem as well.

The above discussion shows that an integer set partition problem can in fact be solved by an approximate solution to the problem for small errors. This also implies that the problem of obtaining an approximate solution to the integer set partition problem can be as hard as finding the exact solution to the problem. In other words, if the allowed errors are less than $1/2n$, the problem of obtaining an approximate solution is equivalent to finding an exact solution to the problem and must also be NP-hard.

We are now ready to show that the problem of obtaining an approximate solution to the distance geometry problem is also NP-hard if the allowed errors for the distances are less than $1/2n$. The argument is based on the fact that the problem of obtaining an approximate solution to the integer set partition problem can be reduced to the problem of obtaining an approximate solution to the distance geometry problem.

Suppose we want to find an approximate solution to an integer set partition problem, with s_j, t_j, and ε_j as defined previously and $\varepsilon_j < 1/2n$ for all j. We construct a 1D distance geometry problem with the following distances:

$$d_{j,j+1} = s_j, \quad 1 \le j \le n,$$
$$d_{1,n+1} = 0. \tag{3.29}$$

Instead of solving this problem directly, we solve it approximately by allowing each distance $d_{j,j+1}$ to have an error $\varepsilon_j < 1/2n$ for all $j = 1, 2, \ldots, n$. Suppose that we have found an approximate solution, x_j, $j = 1, 2, \ldots, n+1$ such that

$$||x_{j+1} - x_j| - d_{j,j+1}| \le \varepsilon_j,$$
$$|x_1 - x_{n+1}| \le \varepsilon_{n+1} \le 1/2. \tag{3.30}$$

Note that

$$\left| \sum_{j=1}^{n} (x_{j+1} - x_j) \right| = |x_{n+1} - x_1| \le 1/2. \tag{3.31}$$

Let $t_j = |x_{j+1} - x_j|$. Then, $x_{j+1} - x_j = +t_j$ or $-t_j$. Let $S_1 = \{j : x_{j+1} - x_j = +t_j\}$ and $S_2 = \{j : x_{j+1} - x_j = -t_j\}$. It follows that

$$\left| \sum_{j=1}^{n} (x_{j+1} - x_j) \right| = \left| \sum_{j \in S_1} (x_{j+1} - x_j) + \sum_{j=S_2} (x_{j+1} - x_j) \right|$$
$$= \left| \sum_{j \in S_1} t_j - \sum_{j=S_2} t_j \right| \le \frac{1}{2}.$$

Along with the fact that $|t_j - s_j| \le \varepsilon_j$ for all $j = 1, 2, \ldots, n$, S_1 and S_2 give a partition for the numbers t_j, $j = 1, 2, \ldots, n$ and hence an approximate solution to the integer set partition problem. This result implies that the problem of obtaining an approximate solution to a distance geometry problem is at least as hard as the problem of finding

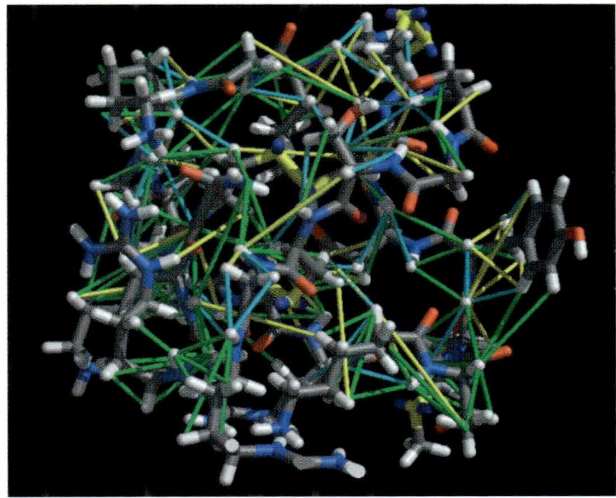

Fig. 3.6. Distance-based modeling. A protein structure can be determined with a set of distance constraints usually obtained from physical experiments or theoretical estimates.

an approximate solution to the integer set partition problem. Since the latter is NP-hard, the former must be as well.

3.3. Distance-based Modeling

A structure can be determined by using different types of data such as the diffraction data obtained from an X-ray crystallographic experiment or the distance data estimated from NMR. By distance-based modeling, we mean a method for structure determination using information on distances between atoms or residues (Fig. 3.6). NMR modeling can certainly be considered as a distance-based method.

3.3.1. *Embedding*

The problem of finding the coordinates of the atoms with a given set of interatomic distances is called embedding. The word derives originally from distance geometry, where embedding can even be considered for general topological spaces. Following this concept, Crippen and Havel (1988) developed an algorithm named embedding, which determines the coordinates of the atoms for a given set of interatomic distances or their ranges. The algorithm has been adopted in software like CNS and X-PLOR, and is widely used in NMR modeling.

The embedding algorithm — as implemented in CNS with certain extensions — starts with a procedure called bound smoothing, continues with another procedure called metrication, and then finally performs the actual embedding. Given a set of distance ranges, the bound smoothing procedure attempts to obtain an estimate of the missing distance ranges. A basic rule for the estimation, for example, is that for the distances among any three atoms i, j, k, if the lower and upper bounds $l_{i,k}$ and $u_{i,k}$ between i and k as well as the bounds $l_{k,j}$ and $u_{k,j}$ between k and j are given, but those $l_{i,j}$ and $u_{i,j}$ between i and j are missing, then

$$u_{i,j} \leq u_{i,k} + u_{k,j},$$
$$l_{i,j} \geq \max\{l_{i,k} - u_{k,j}, l_{k,j} - u_{i,k}\}. \tag{3.32}$$

Once the bounds for all of the distances are estimated, a distance is randomly generated between each pair of bounds. The generated distances may not necessarily be consistent, i.e. they may not even satisfy the triangle inequality. Thus, the metrication procedure is applied to correct the errors when they occur, for example, by regenerating a distance between the corresponding bounds.

After metrication, a SVD algorithm is applied to the generated distances to obtain a set of coordinates for the atoms. If a rank 3 or less decomposition is obtained, the coordinates form a structure whose interatomic distances must satisfy all of the estimated distance ranges. Then, a plausible structure is generated. Otherwise, the coordinates provide a structure that can be considered as an approximation to the true structure. This structure can be refined by using optimization, which is typically done by an energy minimization procedure with the distance ranges as the constraints.

3.3.2. *Least squares method*

For a given set of distances or distance ranges, a more straightforward method to find the coordinates of the atoms is to minimize the total distance errors. The method can be implemented through the solution to a least squares problem for the distances. For example, given a set of distances $d_{i,j}$, for all (i, j) in S, the coordinates $x = \{x_1, x_2, \ldots, x_n\}$ of the atoms can be obtained by solving the following problem:

$$\min \sum_{(i,j)\in S} (||x_i - x_j||^2 - d_{i,j}^2)^2. \tag{3.33}$$

If a set of distance ranges is given instead, a similar least squares formula can also be obtained so that the sum of the squares of the errors is minimized when a structure can be found to fit in all of the distance ranges:

$$\min \sum_{(i,j)\in S} (||x_i - x_j||^2 - u_{i,j}^2)_+^2 + (l_{i,j}^2 - ||x_i - x_j||^2)_+^2, \tag{3.34}$$

where for any function g, $g_+ = g$ when $g > 0$ and $g_+ = 0$ otherwise.

Other formulations similar to formulas (3.33) and (3.34) have also been used, which simply remove the squares on the distances. Therefore, for exact distances, the problem becomes

$$\min \sum_{(i,j)\in S} (||x_i - x_j|| - d_{i,j})^2, \tag{3.35}$$

and for distance bounds,

$$\min \sum_{(i,j)\in S} (||x_i - x_j|| - u_{i,j})_+^2 + (l_{i,j} - ||x_i - x_j||)_+^2. \tag{3.36}$$

The objective functions (3.35) and (3.36) calculate the errors of the distances directly rather than the errors of the squares of the distances, and therefore may be numerically more stable. However, they are not continuously differentiable, and need to be treated with caution when minimized with a conventional optimization method such as the steepest descent direction method, which requires the continuous differentiability of the objective function to converge.

In any case, the advantage of using least squares formulation for the solution of the distance geometry problem is that it does not require all of the distances; in other words, it does not require estimating the missing distances. This avoids not only introducing additional possible errors but also overdetermining the solution, because the determination of the coordinates does not necessarily require all of the distances. Note also that for the above least squares problem, the global minimum of the objective function is known to be zero when the solution to the problem exists. Therefore, in contrast to other global optimization problems, the global minimum for the least squares formulation of the distance geometry problem can be verified relatively easily. Nevertheless, the global minimum is generally still difficult to achieve because the objective function is highly nonconvex and has many local minima.

3.3.3. *Geometric buildup*

Recently, Dong and Wu (2002) have developed a more efficient algorithm called the geometric buildup algorithm for the solution of the distance geometry problem. In an ideal case where all of the required distances are available, the algorithm can find a solution to the problem in $O(n)$ floating-point operations. Given an arbitrary set of distances, the algorithm first finds four atoms that are not in the same plane and determines the coordinates for the four atoms using, say, the SVD method just described, with all of the distances among them (assuming they are available). Then, for any undetermined atom j, the algorithm repeatedly performs a procedure as follows: Find four determined atoms that are not in the same plane and whose distances to atom j are available, and determine the coordinates for atom j. Let $x_i = (x_{i,1}, x_{i,2}, x_{i,3})^{\mathrm{T}}$, $i = 1, 2, 3, 4$, be the coordinate vectors of the four atoms. Then, the coordinates $x_j = (x_{j,1}, x_{j,2}, x_{j,3})^{\mathrm{T}}$ for atom j can be determined by using the distances $d_{i,j}$ from atoms $i = 1, 2, 3, 4$ to atom j. Indeed, x_j can be obtained from the solution of the following system of equations:

$$||x_i||^2 - 2x_i^{\mathrm{T}} x_j + ||x_j||^2 = d_{i,j}^2, \quad i = 1, 2, 3, 4, \tag{3.37}$$

where x_i and $d_{i,j}$, $i = 1, 2, 3, 4$ are known. By subtracting equation i from equation $i + 1$ for $i = 1, 2, 3$, we can eliminate the quadratic terms for x_j to obtain

$$-2(x_{i+1} - x_i)^{\mathrm{T}} x_j = (d_{i+1,j}^2 - d_{i,j}^2) - (||x_{i+1}||^2 - ||x_i||^2), \quad i = 1, 2, 3. \tag{3.38}$$

Let A be a matrix and b a vector:

$$A = -2 \begin{bmatrix} (x_2 - x_1)^{\mathrm{T}} \\ (x_3 - x_2)^{\mathrm{T}} \\ (x_4 - x_3)^{\mathrm{T}} \end{bmatrix},$$

$$b = \begin{bmatrix} (d_{2,j}^2 - d_{1,j}^2) - (||x_2||^2 - ||x_1||^2) \\ (d_{3,j}^2 - d_{2,j}^2) - (||x_3||^2 - ||x_2||^2) \\ (d_{4,j}^2 - d_{3,j}^2) - (||x_4||^2 - ||x_3||^2) \end{bmatrix}.$$

$$(3.39)$$

We then have $Ax_j = b$. Since x_1, x_2, x_3, x_4 are not in the same plane, A must be nonsingular, and we can therefore solve the linear system to obtain a unique solution for x_j. Here, solving the linear system requires only constant time. Since we only need to solve $n - 4$ such systems for $n - 4$ coordinate vectors x_j, the total computation time is proportional to n if, at every step, the required coordinates x_i and distances $d_{i,j}$, $i = 1, 2, 3, 4$ are always available.

The advantage of using the geometric buildup algorithm is that it not only is more efficient than the SVD method, but also requires a smaller set of distances and is easier to extend to problems with sparse sets of distances. The SVD method, on the other hand, requires all of the distances, but it can be used to obtain a good approximate solution to the problem if the distances contain some errors or are not consistent. For such a case, the solution found using the geometric buildup algorithm may or may not be a good approximate, depending on the choice of the four base atoms and hence the distances used to build the structure.

3.3.4. *Potential energy minimization*

Regardless of the methods used to solve a distance geometry problem, the structure may satisfy the given distance constraints, but not the energy requirement. In general, the folded protein should have the lowest possible potential energy. In order for the structure to be energetically favorable, potential energy minimization is often necessary after geometric embedding; sometimes, it may even be helpful for further improving the structure to satisfy additional distance constraints.

Potential energy minimization is usually performed either with some local optimization methods such as the Newton method, the conjugate gradient method, or the steepest-descent direction method, or with a global optimization method such as simulated annealing. The latter of course cannot always provide a global minimum, but may at least find a relatively "good" local minimum. This should be more powerful than a local optimization method, but is in general more expensive computationally.

The potential energy function is usually provided by some conventional software. Several software packages are available, including CHARMM, AMBER, and GROMOS. CNS and X-PLOR also have built-in routines for energy calculation and minimization. The energy functions are defined with the same parameters as in CHARMM. The CHARMM energy function, for example, typically has the following form:

$$
\begin{aligned}
E = &\sum_{\text{bonds}} k_b (b - b_0)^2 + \sum_{\text{angles}} k_\theta (\theta - \theta_0)^2 \\
&+ \sum_{\text{dihedrals}} k_\phi (1 + \cos(n\phi - \delta)) \\
&+ \sum_{\text{impropers}} k_\omega (\omega - \omega_0)^2 + \sum_{\text{Urey–Bradley}} k_u (u - u_0)^2 \\
&+ \sum_{\text{nonbonded}} \varepsilon_{i,j} [(R_{i,j}^{\min}/r_{i,j})^{12} - (R_{i,j}^{\min}/r_{i,j})^6] + q_i q_j / \varepsilon r_{i,j},
\end{aligned}
$$

$$(3.40)$$

where the first term in the function accounts for the energy of the bond stretches, where k_b is the bond force constant and $(b - b_0)$ is the distance from equilibrium that the atoms have moved; the second term in the equation accounts for the bond angles, where k_θ is the angle force constant and $(\theta - \theta_0)$ is the angle from equilibrium between three bonded atoms; the third term is for the dihedral angles, where k_ϕ is the dihedral force constant, n is the multiplicity of the function, ϕ is the dihedral angle, and δ is the phase shift; the fourth term is for the improper angles that are out-of-plane bending, where

k_ω is the force constant and $(\omega - \omega_0)$ is the out-of-plane angle; the Urey–Bradley component comprises the fifth term, where k_u is the respective force constant and u is the distance between the 1,3 atoms in the harmonic potential; and nonbonded interactions between (i,j) pairs of atoms are represented by the last two terms. By definition, the nonbonded forces are only applied to atom pairs separated by at least three bonds. The van der Waals energy is calculated with a standard 12-6 Lennard–Jones potential, and the electrostatic energy with a Coulomb potential. In the Lennard–Jones potential, the $R^{\min}$ term is not the minimum of the potential, but rather where the Lennard–Jones potential crosses the x-axis.

3.4. Structural Analysis

With a set of distance ranges generated from NMR experiments, an ensemble of structures may be determined, all satisfying the given distance constraints. The variation of structures in the ensemble reflects in a certain sense the fluctuation of the protein in solution (Fig. 3.7).

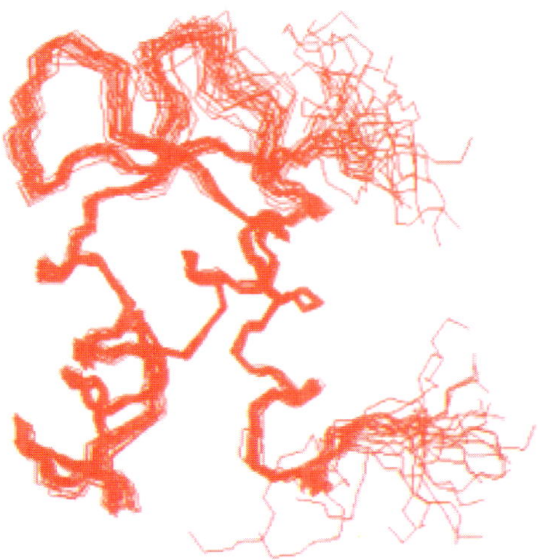

Fig. 3.7. NMR structural ensemble. An ensemble of structures can usually be obtained from NMR showing fluctuations of structures in solution.

It is therefore important to analyze the structures and find their deviations in the ensemble. Given the limited amount of data and possible experimental errors, the determined structures may not be completely correct; and the question of how to evaluate the structures without further experimental evidence also remains a critical issue in practice.

3.4.1. *The coordinate root mean square deviation*

The coordinate root mean square deviation (RMSD) has been widely used for comparing and validating protein structures. It has also been an important tool for structural classification, motif recognition, and structure prediction, where a large number of different proteins must be aligned and compared. Let $X = \{x_i, i = 1, 2, \ldots, n\}$ and $Y = \{y_i, i = 1, 2, \ldots, n\}$ be two $n \times 3$ coordinate matrices for two lists of atoms in proteins A and B, respectively, where $x_i = (x_{i,1}, x_{i,2}, x_{i,3})^{\mathrm{T}}$ is the coordinate vector of the ith atom selected from protein A to be compared with $y_i = (y_{i,1}, y_{i,2}, y_{i,3})^{\mathrm{T}}$, the coordinate vector of the ith atom selected from protein B. Assume that X and Y have been translated so that their centers of geometry are located at the same position, say, at the origin. Then, the structural similarity between the two proteins can be measured by using the coordinate RMSD of the structures as defined by the following:

$$\mathrm{RMSD}(X, Y) = \min_{Q} ||X - YQ||_{\mathrm{F}}/\sqrt{n}, \qquad (3.41)$$

where Q is a 3×3 rotation matrix and $QQ^{\mathrm{T}} = I$, and $|| \cdot ||_{\mathrm{F}}$ is the matrix Frobenius norm. Based on this definition, the RMSD of two structures X and Y is essentially the smallest average coordinate error of the structures for all possible rotations Q of structure Y to fit structure X (Fig. 3.8). Note that X and Y may be the coordinate matrices for the same ($A = B$) or different ($A \neq B$) proteins, and therefore each pair of corresponding atoms need not be of the same type (when $A \neq B$). However, the number of atoms selected to be compared must be the same for A and B (# rows of $X = $ # rows of Y).

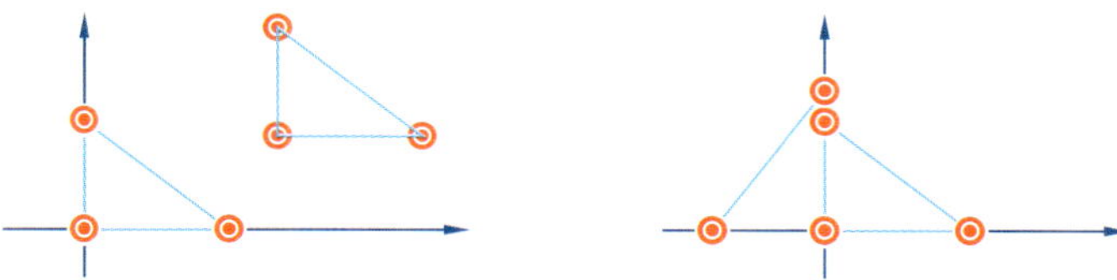

Fig. 3.8. Translation and rotation. The coordinate root mean square deviation (RMSD) can be calculated only after aligning the structures with proper translation and rotation.

In any case, the RMSD calculation requires the solution of an optimization problem, as suggested in its definition. The optimization problem is not trivial to solve if a conventional optimization method is to be used (such as the Newton or steepest descent method). Fortunately, an analytical solution to the problem can actually be obtained with some simple linear algebraic calculations as follows.

Note that

$$||X - YQ||_F^2 = \text{trace}(X^T X) + \text{trace}(Y^T Y) - 2\text{trace}(Q^T Y^T X).$$

Therefore, minimizing the square of $||X - YQ||_F$ is equivalent to maximizing $\text{trace}(Q^T Y^T X)$. Let $C = Y^T X$. Let $C = U\Sigma V^T$ be the SVD of C. Then,

$$\text{trace}(Q^T Y^T X) = \text{trace}(V^T Q^T U\Sigma) \leq \text{trace}(\Sigma).$$

It follows that $Q = UV^T$ maximizes $\text{trace}(Q^T Y^T X)$ and therefore minimizes the square of $||X - YQ||_F$.

3.4.2. *NMR structure evaluation*

The precision of an ensemble of structures determined by NMR is usually measured by the RMSD values of the structures in the ensemble compared with the average structure of the ensemble, in particular by the mean and standard deviation of these values. Let S_E be an ensemble of m structures $X_1, X_2, \ldots, X_m$. Let X_a be the average structure, $X_a = (X_1 + X_2 + \cdots + X_m)/m$. Then, the precision of the structures in S_E can be defined as

$$\text{RMSD\_PRE}(S_E) = \left[\sum_{i=1}^{m} \text{RMSD}(X_i, X_a) \right] \Big/ m. \qquad (3.42)$$

The ensemble RMSD can be used to see how the structure fluctuates. The larger the ensemble RMSD value is, the more flexible the structure must be. On the other hand, if the distance constraints are not tight enough due to experimental errors, the ensemble RMSD may also be able to show how consistent the structures are in the ensemble. The smaller the ensemble RMSD value, the more convergent the structures may be. The proper interpretation of an ensemble RMSD value is a matter of case-by-case analysis. In any event, the value may often be overestimated, since the ensemble of structures determined by current modeling software may not necessarily contain the whole range of structures that are determined by the given distance constraints.

The accuracy of an ensemble of structures determined by NMR is often referred to as the average RMSD value of the structures in the ensemble compared with a reference structure such as an X-ray structure of the protein. Let S_E be an ensemble of m structures $X_1, X_2, \ldots, X_m$. Let X_r be the reference structure. Then, the accuracy of the structures in S_E can be defined as

$$\text{RMSD\_ACC}(S_E) = \left[\sum_{i=1}^{m} \text{RMSD}(X_i, X_r) \right] \Big/ m. \qquad (3.43)$$

The accuracy may be biased toward the reference structure. For example, an NMR structure may not necessarily agree with the X-ray structure, because the former is determined in a more flexible environment and is more of an average shot of the structure moving dynamically in its natural environment, while the latter is determined more statically when the structure is in a fixed crystal state. In any event, the RMSD values of the structures against a reference structure can still provide some valuable information about how the structures fluctuate around or deviate from a regular structure. Their true physical implication is of course subject to further analysis and interpretation, and depends on specific cases.

3.4.3. *Ramachandran plot*

Another commonly used tool for structural evaluation is the Ramachandran plot. The Ramachandran plot is a 2D graph in the ϕ–ψ plane, where ϕ and ψ are the torsion angles around the two flexible bonds joined at C_α in each residue, one between N and C_α and the other between C_α and C'. The plot has four major regions: most favorable, additionally allowed, generously allowed, and disallowed, indicating the preferences of ϕ and ψ angles for protein residues (see Fig. 3.9). The ϕ–ψ angles formed within a residue can be represented by a point on the Ramachandran plot. If the point is in a particular region, we simply say that the corresponding residue is in that region. Usually, a well-determined structure has a high percentage of residues in the most favorable region of the plot. For an ensemble of structures determined by NMR, the average properties of the Ramachandran

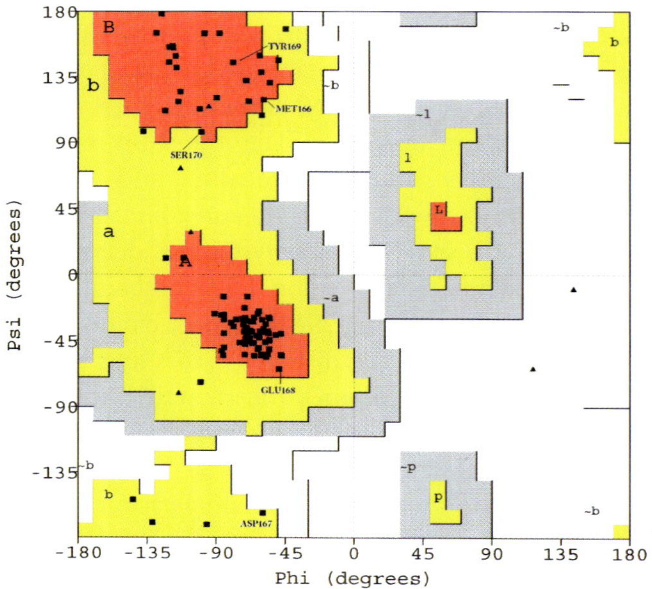

Fig. 3.9. Ramachandran plot. The distribution of protein backbone phi–psi (ϕ–ψ) angles in known protein structures: red — most favorable; yellow — additionally allowed; gray — generously allowed; and white — disallowed.

plots of the structures may be extracted and used for the evaluation of the structures. In addition, for each individual residue, the distribution of ϕ–ψ angles within the ensemble may provide some important insights into how the residue fluctuates in terms of changes in these angles.

3.4.4. *Structure refinement*

Many software packages are available for structure refinement. Take CNS as an example. CNS refers to Crystallography and NMR System, and was developed by Brünger *et al.* (1998). It has been widely used for refining either X-ray or NMR structures. When used for NMR structure refinement, it requires four steps: connectivity calculation, template generation, annealing, and acceptance test. The input data include a set of dihedral angles estimated through *J*-coupling experiments and a set of distance ranges measured in NOE experiments.

Connectivity calculation takes the protein sequence as the input and produces a connectivity file for the backbone of the protein. Template generation uses the connectivity file to construct an extended structure (or a group of extended structures) for the protein as the initial structures for annealing. The annealing process has two options: one with simple simulated annealing, and another with distance geometry combined with simulated annealing. The latter embeds the structure into 3D space by satisfying the distance constraints before performing simulated annealing. Here, embedding solves a distance geometry problem with the NMR distance data, and simulated annealing further minimizes the distance errors and the energy. The last step, acceptance test, evaluates the structures with a group of acceptance criteria including the satisfaction of various experimental constraints and stereochemical requirements.

The refinement procedure is repeated or performed with multiple starting structures, so that an ensemble of structures, all satisfying the NMR distance constraints, can be found. The structures are then evaluated for precision and accuracy. As standard outputs, the average energy values as well as violations of the experimental constraints are calculated, with the energy values separated into different categories including overall energy, bond length, bond angle, van der Waals energy, electrostatic energy, etc. These values can be used to evaluate and compare structures.

Selected Further Readings

Nuclear magnetic resonance

Wüthrich K, *NMR of Proteins and Nucleic Acids*, Wiley, 1986.
Wüthrich K, *NMR in Structural Biology*, World Scientific Publishing Co., 1995.
Gunther H, *NMR Spectroscopy: Basic Principles, Concepts, and Applications in Chemistry*, Wiley, 2001.
Keeler J, *Understanding NMR Spectroscopy*, Wiley, 2005.
Cavanagh J, Fairbrother III WJ, Palmer AG, Skelton NJ, *Protein NMR Spectroscopy: Principles and Practice*, Academic Press, 2006.

Distance geometry

Blumenthal LM, *Theory and Applications of Distance Geometry*, Oxford Clarendon Press, 1953.
Torgerson WS, *Theory and Method of Scaling*, Wiley, 1958.
Kruskal JB, Non-metric multidimensional scaling: A numerical method, *Psychometrika* **29**: 115–129, 1964.
Guttman L, A general non-metric technique for finding the smallest coordinate space for a configuration of points, *Psychometrika* **33**: 469–506, 1968.
Saxe JB, Embeddability of weighted graphs in k-space is strongly NP-hard, in *Proc 17th Allerton Conference in Communications, Control and Computing*, pp. 480–489, 1979.
De Leeuw J, Heiser W, Multidimensional scaling with restrictions on configuration, in *Multivariate Analysis*, Krishnaiah VPR (ed.), North-Holland, pp. 501–522, 1980.
Crippen GM, Havel TF, *Distance Geometry and Molecular Conformation*, Wiley, 1988.
Havel TF, Distance geometry, in *Encyclopedia of Nuclear Magnetic Resonance*, Grant DM, Harris RK (eds.), Wiley, pp. 1701–1710, 1995.

Havel T, An evaluation of computational strategies for use in the determination of protein structure from distance constraints obtained by nuclear magnetic resonance, *Prog Biophys Mol Biol* **56**: 43–78, 1991.

Kuntz ID, Thomason JF, Oshiro CM, Distance geometry, *Methods Enzymol* **177**: 159–204, 1993.

Glunt W, Hayden TL, Liu WM, The embedding problem for predistance matrices, *Bull Math Biol* **53**: 769–796, 1991.

Glunt W, Hayden TL, Hong S, Wells J, An alternating projection algorithm for computing the nearest Euclidean distance matrix, *SIAM J Matrix Anal Appl* **11**: 589–600, 1990.

Glunt W, Hayden TL, Raydan M, Molecular conformations from distance matrices, *J Comput Chem* **14**: 114–120, 1993.

Hayden TL, Wells J, Liu W, Tarazaga P, The cone of Euclidean distance matrices, *Linear Algebra Appl* **144**: 153–169, 1991.

Hendrickson A, Conditions for unique graph realizations, *SIAM J Comput* **21**: 65–84, 1992.

Hendrickson A, The molecule problem: Exploiting structure in global optimization, *SIAM J Optim* **5**: 835–857, 1995.

Moré J, Wu Z, ε-Optimal solutions to distance geometry problems via global continuation, in *Global Minimization of Non-convex Energy Functions: Molecular Conformation and Protein Folding*, Pardalos PM, Shalloway D, Xue G (eds.), American Mathematical Society, pp. 151–168, 1996.

Moré J, Wu Z, Global continuation for distance geometry problems, *SIAM J Optim* **7**: 814–836, 1997.

Moré, J, Wu Z, Distance geometry optimization for protein structures, *J Glob Optim* **15**: 219–234, 1999.

Kearsly A, Tapia R, Trosset M, Solution of the metric STRESS and SSTRESS problems in multidimensional scaling by Newton's method, *Comput Stat* **13**: 369–396, 1998.

Trosset M, Applications of multidimensional scaling to molecular conformation, *Comput Sci Stat* **29**: 148–152, 1998.

So AM, Ye Y, Theory of semi-definite programming for sensor network localization, *Math Program* **109**: 367–384, 2007.

Biswas P, Liang T, Toh K, Ye Y, An SDP based approach for anchor-free 3D graph realization, Technical report, Department of Management Science and Engineering and Department of Electrical Engineering, Stanford University, Stanford, CA, 2007.

Dong Q, Wu Z, A linear-time algorithm for solving the molecular distance geometry problem with exact inter-atomic distances, *J Global Optim* **22**: 365–375, 2002.

Dong Q, Wu Z, A geometric buildup algorithm for solving the molecular distance geometry problem with sparse distance data, *J Global Optim* **26**: 321–333, 2003.

Wu D, Wu Z, An updated geometric buildup algorithm for solving the molecular distance geometry problem with sparse distance data, *J Global Optim* **37**: 661–673, 2007.

Distance-based protein modeling

Brünger AT, Niles M, Computational challenges for macromolecular modeling, in *Reviews in Computational Chemistry*, Vol. 5, Lipkowitz KB, Boyd DB (eds.), VCH Publishers, pp. 299–335, 1993.

Brünger AT, Adams PD, Clore GM, DeLano WL, Gros P, Grosse-Kunstleve RW, Jiang S, Kuszewski J, Nilges N, Pannu NS, Read RJ, Rice LM, Simonson T, Warren GL, Crystallography and NMR system: A new software suite for macromolecular structure determination, *Acta Cryst* **D54**: 905–921, 1998.

Kuszewski J, Niles M, Brünger AT, Sampling and efficiency of metric matrix distance geometry: A novel partial metrization algorithm, *J Biomol NMR* **2**: 33–56, 1992.

Clore GM, Robien MA, Gronenborn AM, Exploring the limits of precision and accuracy of protein structures determined by nuclear magnetic resonance spectroscopy, *J Mol Biol* **231**: 82–102, 1993.

Tjandra N, Omichinski JG, Gronenborn AM, Clore GM, Bax A, Use of dipolar ^{1}H-^{15}N and ^{1}H-^{13}C couplings in the structure determination of magnetically oriented macromolecules in solution, *Nat Struct Biol* **4**: 732–738, 1997.

Clore GM, Gronenborn AM, Tjandra N, Direct structure refinement against residual dipolar couplings in the presence of rhombicity of unknown magnitude, *J Magn Reson* **131**: 159–162, 1998.

Grishaev A, Bax A, An empirical backbone-backbone hydrogen-bonding potential in proteins and its applications to NMR structure refinement and validation, *J Am Chem Soc* **126**: 7281–7292, 2004.

Doreleijers JF, Mading S, Maziuk D, Sojourner K, Yin L, Zhu J, Markley JL, Ulrich EL, BioMagResBank database with sets of experimental NMR constraints corresponding to the structures of over 1400 biomolecules deposited in the Protein Data Bank, *J Biomol NMR* **26**: 139–146, 2003.

Havel TF, Snow ME, A new method for building protein conformations from sequence alignments with homologues of known structure, *J Mol Biol* **217**: 1–7, 1991.

Srinivasan S, March CJ, Sudarsanam S, An automated method for modeling proteins on known templates using distance geometry, *Protein Sci* **2**: 277–289, 1993.

Huang ES, Samudrala R, Ponder JW, Distance geometry generates native-like folds for small helical proteins using the consensus distances of predicted protein structures, *Protein Sci* **7**: 1998–2003, 1998.

Structural analysis

Morris AL, MacArthur MW, Hutchinson EG, Thornton JM, Stereochemical quality of protein structure coordinates, *Proteins* **12**: 345–364, 1992.

Liu Y, Zhao D, Altman R, Jardetzky O, A systematic comparison of three structure determination methods from NMR data: Dependence upon quality and quantity of data, *J Biomol NMR* **2**: 373–388, 1992.

Wagner G, Hyberts SG, Havel TF, NMR structure determination in solution: A critique and comparison with X-ray crystallography, *Annu Rev Biophys Biomol Struct* **21**: 167–198, 1992.

Hooft RW, Vriend G, Sander C, Abola EE, Errors in protein structures, *Nature* **381**: 272, 1996.

Doreleijers JF, Rullmann JA, Kaptein R, Quality assessment of NMR structures: A statistical survey, *J Mol Biol* **281**: 149–164, 1998.

Doreleijers JF, Raves ML, Rullmann T, Kaptein R, Completeness of NOEs in protein structure: A statistical analysis of NMR, *J Biomol NMR* **14**: 123–132, 1999.

Doreleijers JF, Vriend G, Raves ML, Kaptein R, Validation of nuclear magnetic resonance structures of proteins and nucleic acids: Hydrogen geometry and nomenclature, *Proteins* **37**: 404–416, 1999.

Garbuzynskiy SO, Melnik BS, Lobanov MY, Finkelstein AV, Galzitskaya OV, Comparison of X-ray and NMR structures: Is there a systematic difference in residue contacts between X-ray- and NMR-resolved protein structures? *Proteins* **60**: 139–147, 2005.

Laskowski RA, MacArthur MW, Moss DS, Thornton JM, PROCHECK: A program to check the stereochemical quality of protein structures, *J Appl Crystallogr* **26**: 283–291, 1993.

Laskowski RA, Rullmannn JA, MacArthur MW, Kaptein R, Thornton JM, AQUA and PROCHECK-NMR: Programs for checking the quality of protein structures solved by NMR, *J Biomol NMR* 8: 477–486, 1996.

Word JM, Lovell SC, LaBean TH, Taylor HC, Zalis ME, Presley BK, Richardson JS, Richardson DC, Visualizing and quantifying molecular goodness-of-fit: Small-probe contact dots with explicit hydrogen atoms, *J Mol Biol* 285: 1711–1733, 1999.

Word JM, Bateman Jr RC, Presley BK, Lovell SC, Richardson DC, Exploring steric constraints on protein mutations using MAGE/PROBE, *Protein Sci* 9: 2251–2259, 2000.

Gronwald W, Kirchhofer R, Gorler A, Kremer W, Ganslmeier B, Neidig KP, Kalbitzer HR, RFAC, a program for automated NMR R-factor estimation, *J Biomol NMR* 17: 137–151, 2000.

Huang YJ, Powers R, Montelione GT, Protein NMR recall, precision, and F-measure scores (RPF scores): Structure quality assessment measures based on information retrieval statistics, *J Am Chem Soc* 127: 1665–1674, 2005.

Nabuurs SB, Spronk CA, Krieger E, Maassen H, Vriend G, Vuister GW, Quantitative evaluation of experimental NMR restraints, *J Am Chem Soc* 125: 12026–12034, 2003.

Spronk CA, Nabuurs SB, Bonvin AM, Krieger E, Vuister GW, Vriend G, The precision of NMR structure ensembles revisited, *J Biomol NMR* 25: 225–234, 2003.

Spronk CA, Nabuurs SB, Krieger E, Vriend G, Vuister GW, Validation of protein structures derived by NMR spectroscopy, *Prog Nucl Magn Reson Spectrosc* 45: 315–337, 2004.

Nabuurs SB, Spronk CA, Vuister GW, Vriend G, Traditional biomolecular structure determination by NMR spectroscopy allows for major errors, *PLoS Comput Biol* 2: 71–79, 2006.

Snyder DA, Bhattacharya A, Huang YJ, Montelione GT, Assessing precision and accuracy of protein structures derived from NMR data, *Proteins* 59: 655–661, 2005.

Snyder DA, Montelione GT, Clustering algorithms for identifying core atom sets and for assessing the precision of protein structure ensembles, *Proteins* 59: 673–686, 2005.

Kuszewski J, Gronenborn AM, Clore GM, Improving the quality of NMR and crystallographic protein structures by means of a conformational database potential derived from structure databases, *Protein Sci* 5: 1067–1080, 1996.

Wall ME, Subramaniam S, Phillips Jr GN, Protein structure determination using a database of interatomic distance probabilities, *Protein Sci* 8: 2720–2727, 1999.

Cui F, Jernigan R, Wu Z, Refinement of NMR-determined protein structures with database derived distance constraints, *J Bioinform Comput Biol* 3: 1315–1329, 2005.

Wu D, Jernigan R, Wu Z, Refinement of NMR-determined protein structures with database derived mean-force potentials, *Proteins* 68: 232–242, 2007.

Chapter 4

Potential Energy Minimization

Anfinsen (1972) showed that the amino acid sequence of a protein determines its three-dimensional (3D) folding. Therefore, it is possible to find the protein structure based on only its sequence information. Along this line, research has been pursued on the determination of protein structure using only the information about what amino acids are in the protein and how they interact physically. A thermodynamic hypothesis on the physical interactions in a protein is that the atoms in the protein interact with each other in order to reach an equilibrium state where the potential energy of the protein is minimized. Based on this hypothesis, the protein native structure can in principle be determined if the potential energy minimum of the protein can be found. The latter problem may be solved through potential energy minimization, if the potential energy function is provided. Of course, this problem is not trivial to solve because the protein native structure corresponds to the global energy minimum, which is difficult to find if there are many local energy minima (Fig. 4.1). The protein potential energy functions are highly nonconvex functions and indeed have many local energy minima.

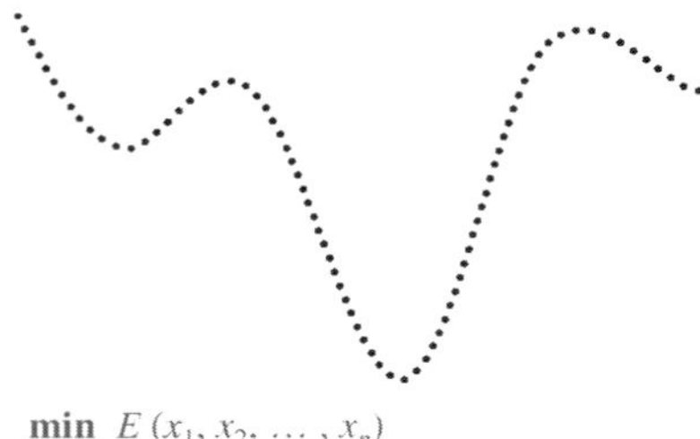

Fig. 4.1. Energy minimization. The native fold of the protein is assumed to correspond to the global minimum of the potential energy and may therefore be determined via global optimization.

4.1. Potential Energy Function

The potential energy of a protein can in principle be evaluated with a quantum chemistry approach, but the required computation is not affordable even on supercomputers. Semiempirical functions have been developed to compute different energy terms separately and approximately, and have proven to be practical.

A correct energy function is vital to the success of the energy minimization approach for folding, for the obvious reason that the search for a global energy minimum may be misguided otherwise. There are software packages that can be used to calculate the potential energy of a protein based on well-developed semiempirical models. However, the functions (or parameters) used in these calculations are still under constant improvement for both efficiency and reliability.

4.1.1. *Quantum chemistry calculation*

From the viewpoint of quantum mechanics, a protein, like any other physical system, can be considered as a system of particles with certain wave properties. Let ψ be the wave function describing the behavior of the entire particle system. Then, ψ satisfies the following wave equation called the time-dependent Schrödinger equation,

$$\hat{H}\psi(x, t) = i\hbar\frac{\partial}{\partial t}\psi(x, t), \qquad (4.1)$$

where $\hat{H}$ is the Hamiltonian operator, and ψ is defined as a function of state variable x and time t. Associated with the above equation is the time-independent Schrödinger equation,

$$\hat{H}\varphi(x) = E\varphi(x), \tag{4.2}$$

where $\hat{H}$ is assumed to be time-independent and φ is a time-independent wave function.

The time-independent Schrödinger equation can be viewed as a generalized eigenvalue equation, where E is the eigenvalue and φ is the eigenfunction. For a single particle of mass m, for example, the Hamiltonian operator $\hat{H}$ can be defined as follows:

$$\hat{H} = -\frac{\hbar^2}{2m}\nabla^2 + P(x), \tag{4.3}$$

where P is the potential of the system. With this operator, the time-independent Schrödinger equation becomes

$$-\frac{\hbar^2}{2m}\nabla^2\varphi(x) + P(x) = E\varphi(x). \tag{4.4}$$

A Schrödinger equation can be solved by making a finite series approximation to the wave function. The series typically consists of a large set of basis functions. By substituting the wave function in the equation with the series approximation, a system of algebraic equations can be obtained, with the parameters in the basis functions as the variables. The basis functions and hence the series approximation to the wave function can then be determined by solving the system of equations. In fact, in many cases, the system of equations reduces to an eigenvalue problem, with the eigenvalues corresponding to the energy levels and the eigenvectors to the approximations of the wave functions.

For a large molecular system, the required series approximation to the wave function can be very complicated in terms of both the number and complexity of the basis functions. As a result, an enormous amount of computation is required to obtain a solution to the reduced eigenvalue problem. Alternative approaches have been proposed for a more efficient solution to the problem, but their computational costs remain too expensive for large molecular systems like proteins.

4.1.2. *Semiempirical approximation*

To avoid the expensive quantum mechanical computation, an atomic-level classical mechanical model has been adopted for the evaluation of the potential energy of a protein. With this model, the potential energy is considered as a result of the interactions among all of the atoms in the protein. Since there are different interactions, the potential energy is divided into different categories such as bond length potentials, bond angle potentials, torsion angle potentials, electrostatic potentials, van der Waals potentials, etc. The total potential energy is equal to the sum of all these potentials (Fig. 4.2).

When two atoms are connected by a chemical bond, they tend to maintain a fixed distance. This interaction between the two atoms can be described by using the bond length potential, which is usually approximated by a quadratic function with the equilibrium bond length at the minimum of the function. Physically, this approximation is equivalent to considering the interaction between two bonded atoms as if there is a spring between them. The coefficient of the function therefore corresponds to a spring constant, which depends on the type of bond and the atoms involved, but can usually be estimated through chemical experiments. This is why such types of potentials are also called semiempirical potentials.

Let S_{bond} be the index set for the pairs of atoms that are connected by a chemical bond, and E_{bond} the total bond energy.

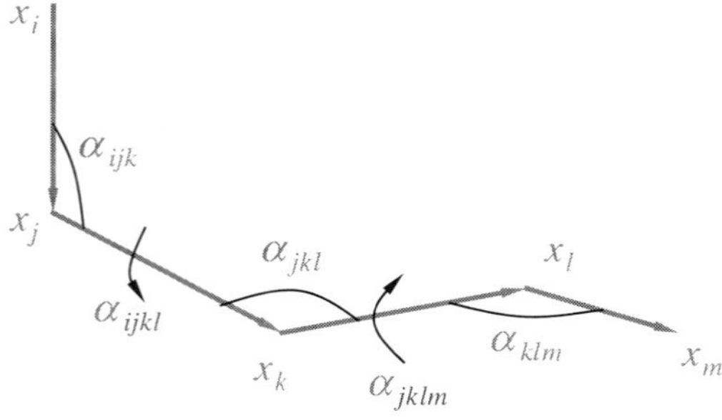

Fig. 4.2. Empirical potentials. Functions can be defined to approximate the potential energies for various atomic-level interactions.

Then,

$$E_{\text{bond}} = \sum_{(i,j)\in S_{\text{bond}}} k_{ij}^{\text{b}}(r_{ij} - r_{ij}^0)^2, \qquad (4.5)$$

where k_{ij}^{b} and r_{ij}^0 are the force constant and the equilibrium length of the bond between atoms i and j, and $r_{ij} = ||x_i - x_j||$ is the current bond length.

Similarly, when three atoms are connected with two chemical bonds, the two bonds tend to form a fixed angle. Such an interaction can be described by using the bond angle potential, which can also be approximated by a quadratic function with the equilibrium bond angle at the minimum of the function. Physically, this implies that the interaction between two neighboring bonds can be considered as if there is a spring holding the two bonds. The coefficient of the function is again a spring constant, which depends on the types of atoms and bonds, and can be estimated empirically.

Let S_{angle} be the index set for the triplet of atoms that are connected by two chemical bonds, and E_{angle} the total bond angle energy. Then,

$$E_{\text{angle}} = \sum_{(i,j,k)\in S_{\text{angle}}} k_{ijk}^{\text{a}}(\alpha_{ijk} - \alpha_{ijk}^0)^2, \qquad (4.6)$$

where k_{ijk}^{a} and α_{ijk}^0 are the force constant and equilibrium angle of the bond angle α_{ijk} between the two bonds that connect atoms i, j, and k.

If four atoms are connected with three bonds, the bond in the middle may be rotated with a certain angle. This rotation, also called the torsion, often prefers specific angles. Such a preference depends on the types of atoms and bonds, but can be described by the torsion angle potential. The latter is often calculated with an empirical cosine function.

Let S_{torsion} be the index set for the quartet of atoms that are connected by three chemical bonds, and E_{torsion} the total torsion angle energy. Then,

$$E_{\text{torsion}} = \sum_{(i,j,k,l)\in S_{\text{torsion}}} k_{ijkl}^{\text{t}}[1 + \cos(n\alpha_{ijkl} - \alpha_{ijkl}^0)], \qquad (4.7)$$

where k^{t}_{ijkl} and α^0_{ijkl} are the force constant and equilibrium angle of the torsion angle α_{ijkl} of the three bonds that connect atoms i, j, k, and l.

The electrostatic potential is used to compute the potential energy between charged atoms. This potential can be approximated by a function inversely proportional to the distance between the two atoms. The coefficient of the function depends on the types of atoms involved, and can be estimated using the electrostatic theory and physical experiments.

Let S_{electro} be the index set for the pairs of atoms with electrostatic interactions, and E_{electro} the total electrostatic energy. Then,

$$E_{\mathrm{electro}} = \sum_{(i,j) \in S_{\mathrm{electro}}} \frac{q_i q_j}{e_{ij} r_{ij}}, \qquad (4.8)$$

where e_{ij} is a constant; q_i and q_j are the charges of atoms i and j, respectively; and $r_{ij} = ||x_i - x_j||$.

Finally, between nonbonded and noncharged atoms, there is still an important interaction known as the van der Waals weak interaction. This interaction also contributes to the total potential energy and cannot be neglected. The van der Waals interaction can be approximated by using a so-called Lennard–Jones potential, which is equal to zero if the two atoms are far apart and to infinity if they are too close, but reaches a local minimum at an appropriate distance. Two parameters must be determined: the minimizing distance and the minimum potential. They are usually estimated partly theoretically based on the chemical theory of the van der Waals radii of the atoms.

Let S_{vdw} be the index set for the pairs of atoms with van der Waals interactions, and E_{vdw} the total van der Waals energy. Then,

$$E_{\mathrm{vdw}} = \sum_{(i,j) \in S_{\mathrm{vdw}}} \varepsilon_{ij} \left[\left(\frac{\sigma_{ij}}{r_{ij}} \right)^{12} - 2 \left(\frac{\sigma_{ij}}{r_{ij}} \right)^{6} \right], \qquad (4.9)$$

where each (i,j) term reaches the minimum ε_{ij} at $r_{ij} = \sigma_{ij}$.

For different types of potentials, different mathematical functions are required, but the functions may also vary with different parameter values for different atom pairs, triplets, quartets, etc. Fortunately, in proteins, there are only a small number of atom types and hence a small number of distinct atom pairs, triplets, quartets, etc. Therefore, the number of possible parameter sets will not be too large and can be precalculated and stored, and applied to the functions whenever they are required. By collecting all of the different types of energy together, the total potential energy of a protein can be obtained. Let E be the total energy of the protein. Then,

$$E = E_{\text{bond}} + E_{\text{angle}} + E_{\text{torsion}} + E_{\text{electro}} + E_{\text{vdw}}. \qquad (4.10)$$

4.1.3. *Protein energy landscape*

The global energy minimum may or may not be difficult to find, depending on the energy landscape. If the energy landscape can be approximated by a convex function, then the global energy minimum can be found in the same way as a local energy minimum. However, if many local energy minima are distributed randomly over the entire space, then it may be impossible to determine the global minimum.

The protein energy landscape is not simply convex in shape; the local energy minima, despite there being many, are not distributed chaotically either. It is believed that the protein energy landscape forms a funnel-like shape, with many local minima around the wall of the funnel and the global minimum located somewhere in the bottom (Fig. 4.3). If this belief is correct, two properties must make the protein energy landscape not so arbitrary. First, the protein energy landscape has only one big funnel, instead of many multiple funnels. Second, the local energy minima around the wall of the funnel must be relatively shallow so that they can be escaped to reach the bottom region around the global energy minimum; otherwise, folding would not be so successful every time as observed in nature.

The funnel theory has provided an overall guideline for the protein energy landscape, but not any specific or detailed information about the local structures. It is still not clear how the global energy minimum can be located. In fact, it is almost impossible to verify if an energy minimum is a global energy minimum, given the complexity and dimensionality of the protein energy landscape. As a result, it

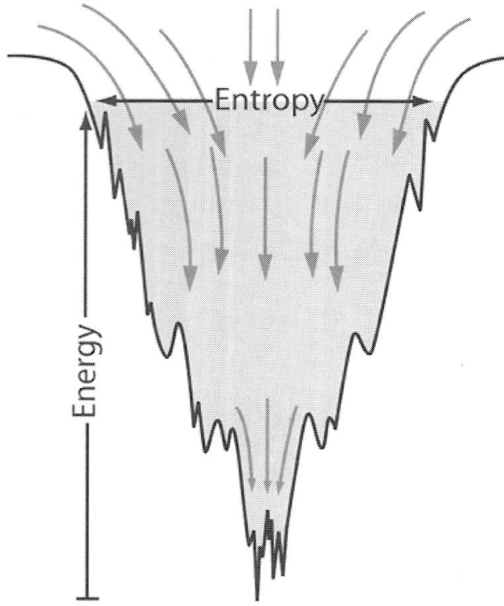

Fig. 4.3. Protein energy landscape. The energy function of a protein may have a funnel shape, with the global minimum at the bottom and many local minima around the wall of the funnel.

is even hard to tell if an experimentally determined structure indeed corresponds to a global energy minimum.

In practice, a known structure sometimes turns out to correspond to a local minimum of the energy function, and is significantly different from the structure that corresponds to the global minimum of the function. In this case, if we believe that the known structure is the native structure and has the least energy, there must be some errors in the energy function. Indeed, the energy functions have to be used with caution, because they are just approximations to the true energy functions and may not include all necessary and correct interactions.

4.1.4. *Implicit and explicit solvent effects*

Solvent effects must also be considered for the proper folding of proteins. The potential energy must therefore include the physical interactions within the protein as well as those between the protein and the solution. There are two ways of including solvent effects: explicit and implicit.

The explicit approach places the water molecules explicitly around the protein, and an empirical potential function is developed to compute the potential energy due to the interactions between the protein and the water molecules. This approach is appropriate when only a small number of water molecules are included. If a large number of water molecules are considered, the calculations may not be accurate using the explicit model.

Instead, the implicit model may provide more accurate results and may also be more efficient to compute in the case of a large number of water molecules. The most commonly used implicit approaches are, for example, using a modified function for electrostatic potential that takes the solvent effects into account, or obtaining the charge density of the solution by solving a Poisson–Boltzmann equation.

4.2. Local Optimization

A general optimization problem is to find, for a given function f, a specific value of x such that f is minimized. If the variable x has to satisfy certain conditions, the problem is called a constrained optimization problem; otherwise, it is called unconstrained. If f is minimized within a small neighborhood of x, $f(x)$ is called a local minimum. If f is minimized for all feasible values of x, $f(x)$ is called a global minimum. The procedure for finding a local minimum is called local optimization, while that for finding a global minimum is known as global optimization.

4.2.1. *Steepest-descent direction method*

Assume that an initial point x^0 is given. We then want to make some changes from x^0 along a certain direction so that the new point x^1 is closer to the optimal solution or, in other words, the function value $f(x^1)$ is lower than $f(x^0)$. We can continue the process iteratively until the function value cannot be further reduced. In order to continuously reduce the function value, at every step, it is desirable for the selected direction to be a descent direction; otherwise, the updated x along that direction may instead cause the function value to increase.

One way of choosing a descent direction is to use the so-called steepest-descent direction. At any point x^k, the steepest-descent direction for a function f is that along the negative gradient direction. Let the gradient of the function at x^k be $\nabla f(x^k)$. The steepest-descent direction p^k for f at x^k would be

$$p^k = -\nabla f(x^k)/||\nabla f(x^k)||. \tag{4.11}$$

A steepest-descent direction method for local optimization uses an iterative formula

$$x^{k+1} = x^k + \alpha^k p^k, \quad k = 0, 1, \ldots \tag{4.12}$$

to generate a sequence of points $\{x^k\}$, where α^k is a step size selected in every iteration so that the new iterate satisfies certain decreasing criteria.

A particular way of choosing α^k is such that $f(x^k + \alpha^k p^k)$ is minimized along a fixed direction p^k. It is called the exact line search. By always choosing an appropriate step size, the sequence of points $\{x^k\}$ generated by the steepest-descent direction method will converge to a local optimal solution to the minimization problem for f. The convergence rate is linear.

4.2.2. *Conjugate gradient method*

The idea of the conjugate gradient method is to use a set of directions that are conjugate to each other with respect to the Hessian of the objective function, when searching for a local minimum of the objective function. Let f be the objective function and ∇f the gradient. Then, a sequence of conjugate directions can be generated by using the following formula:

$$
\begin{aligned}
p^0 &= -\nabla f(x^0), \\
p^{k+1} &= -\nabla f(x^{k+1}) + \beta^{k+1} p^k, \quad k > 0,
\end{aligned}
\tag{4.13}
$$

where

$$x^{k+1} = x^k + \alpha^k p^k,$$

$$\beta^{k+1} = [\nabla f(x^{k+1})^\mathrm{T} \nabla f(x^{k+1})]/[\nabla f(x^k)^\mathrm{T} \nabla f(x^k)].$$

Here, since the directions are generated with the gradients, they are called the conjugate gradient directions, and the method is therefore named the conjugate gradient method.

The advantage of using the conjugate gradient method is that the method has a good chance of converging in a finite number of steps, if the objective function is a quadratic or almost quadratic function. In fact, we can show that if the objective function is quadratic and the Hessian is positive definite, and if the exact line search is used to generate the iterates, then the conjugate gradient method can converge to a local optimal solution to the minimization problem for the given objective function in less than or equal to n steps, where n is the dimension of the problem.

4.2.3. *Newton method*

The Newton method uses another different search direction. Let f be the objective function, ∇f the gradient, and $\nabla^2 f$ the Hessian. Then, the Newton direction at any point x^k is

$$p^k = -[\nabla^2 f(x^k)]^{-1} \nabla f(x^k). \tag{4.14}$$

The Newton direction is not necessarily a descent direction, unless it is very close to a local optimal solution. Even if it is a descent direction, an appropriate step size is still required. Therefore, the update formula is the same as for other methods:

$$x^{k+1} = x^k + \alpha^k p^k, \quad k = 0, 1, \ldots. \tag{4.15}$$

The sequence of iterates generated by the Newton method converges to a local optimal solution to the minimization problem for the given objective function, if the initial point is close enough to the solution. The convergence rate is quadratic, and therefore it converges much faster than linearly converging methods. However, the computation of the Newton step is much more expensive, since it requires the inverse of the Hessian. It is not stable to directly compute the inverse of the matrix. Usually, it is done by solving a linear system,

$$\nabla^2 f(x^k) p^k = -\nabla f(x^k), \tag{4.16}$$

to obtain the Newton direction p^k (as the multiplication of the negative gradient of the objective function with the inverse of the Hessian).

In general, the steepest-descent direction takes an order of n floating-point computations, then the conjugate gradient direction takes an order of n^2, and the Newton direction takes an order of n^3. However, in terms of the convergence rates of the methods, they are preferred in the order of the Newton method, the conjugate gradient method, and the steepest-descent direction method.

4.2.4. *The quasi-Newton method*

The motivation for developing the quasi-Newton method was to reduce the computational cost while still maintaining the fast convergence rate of the Newton method. Since the most expensive part of the Newton method is the computation of the Newton direction, for which the solution of a linear system of equations is required, the quasi-Newton method approximates the Hessian in order to reduce the cost of computing the Newton direction.

For this purpose, at every step, the Hessian is obtained approximately by making a low-rank update to the previous Hessian. By doing so, the inverse of the Hessian can easily be obtained and, in particular, the Newton direction can be obtained in an order of n^2 floating-point operations. Meanwhile, the convergence rate of the quasi-Newton method can be kept superlinear, which is almost quadratic when close to the optimal solution.

Several formulas can be used to update the Hessian, including the symmetric rank-one update (SR1), the Davidon-Fletcher-Powell (DFP), and the Broyden-Fletcher-Goldfarb-Shanno (BFGS) formulas. We list some commonly used formulas below. We use B_k to represent the approximation to the Hessian in the kth iteration, H_k the approximation to the inverse of the Hessian, and

$$
\begin{aligned}
s_k &= x^{k+1} - x^k, \\
y_k &= \nabla f(x^{k+1}) - \nabla f(x^k), \\
\rho_k &= 1/y_k^\mathrm{T} s_k.
\end{aligned}
$$

Note that each update scheme has two formulas, one for B_k and another for H_k. The latter is called the inverse update. Note also that SR1 requires symmetry of the matrix, and BFGS requires the matrix to be symmetric and positive definite.

$$
\text{(SR1)} \quad B_{k+1} = B_k + \frac{(y_k - B_k s_k)(y_k - B_k s_k)}{(y_k - B_k s_k)^\mathrm{T} s_k}, \tag{4.17}
$$

$$
\text{(SR1)} \quad H_{k+1} = H_k + \frac{(s_k - H_k y_k)(s_k - H_k y_k)}{(s_k - H_k y_k)^\mathrm{T} y_k}; \tag{4.18}
$$

$$
\text{(DFP)} \quad B_{k+1} = (I - \rho_k y_k s_k^\mathrm{T}) B_k (I - \rho_k s_k y_k^\mathrm{T}) + \rho_k y_k y_k^\mathrm{T}, \tag{4.19}
$$

$$
\text{(DFP)} \quad H_{k+1} = H_k - \frac{H_k y_k y_k^\mathrm{T} H_k}{y_k^\mathrm{T} H_k y_k} + \frac{s_k s_k^\mathrm{T}}{y_k^\mathrm{T} s_k}; \tag{4.20}
$$

$$
\text{(BFGS)} \quad B_{k+1} = B_k - \frac{B_k s_k s_k^\mathrm{T} B_k}{s_k^\mathrm{T} B_k s_k} + \frac{y_k y_k^\mathrm{T}}{y_k^\mathrm{T} s_k}, \tag{4.21}
$$

$$
\text{(BFGS)} \quad H_{k+1} = (I - \rho_k s_k y_k^\mathrm{T}) H_k (I - \rho_k y_k s_k^\mathrm{T}) + \rho_k s_k s_k^\mathrm{T}. \tag{4.22}
$$

4.3. Global Optimization

In general, a global minimum is difficult to determine. First, it is even difficult to know if a given minimum is indeed a global minimum, unless the entire function domain is examined, which is impossible for a function of more than a few variables. Second, a general non-convex function can vary arbitrarily, and there are no general rules for progressively searching for a global minimum. For example, if we always follow a descent direction, we will eventually be trapped in a local minimum. At a local minimum, if we try to jump out along a certain direction, we may end up with another local minimum; and if we repeat, the chance of reaching a global minimum will not necessarily improve.

4.3.1. *Multi-start method*

The simplest method for global optimization is perhaps the multi-start method. The idea is to first generate a set of initial points evenly distributed in the function domain, and then start with these initial points to obtain a set of local minima (by using any local optimization method), among which the best is selected as the candidate for the global minimum.

Naturally, if sufficiently many initial points are generated and if the local minima are also relatively evenly distributed in the function domain, the chance of finding a global minimum using such a multi-start scheme is rather high. The more initial points generated, the higher the chance of finding the global minimum, but of course the more costly the method will be computationally.

However, if the local minima are not evenly distributed, which is most likely the case in practice, the multi-start method will have a great chance of missing the global minimum even if a large number of initial points are used. Also, in high dimension, the number of initial points required for a complete coverage of the function domain will be too large, and the follow-up local optimization will be too much to finish.

In any case, the multi-start method performs reasonably well in practice. It is easy to implement and can therefore be used as a benchmark method for global optimization. The truth is that so far, although hundreds of global optimization algorithms have been proposed, not so many have always outperformed the multi-start method for arbitrarily given test cases.

4.3.2. *Stochastic search*

Stochastic search can be considered as a smarter way of performing multi-start sampling. A popular stochastic search scheme is the so-called multi-level single-linkage method. This method has two stages. First, a large set of points is sampled randomly in the domain of the objective function. A subset of points is selected as the starting points for local optimization. Second, local optimization is applied to the selected points to obtain a set of local minima. The process continues iteratively until a certain stopping condition is satisfied.

Critical to the multi-level single-linkage method are the strategy for selecting the starting points for local optimization and the rule to stop the iteration. First, the objective function is evaluated at the sampled points. Then, only the points that were not used previously as starting points can be selected. The points should also be those with the lowest function values within their small neighborhood. More accurately, let x be such a point and f the objective function. Then,

$$f(x) \leq f(y), \quad \text{for all generated } y, \quad ||y - x|| < r(k), \qquad (4.23)$$

where $r(k)$ is the neighborhood size at step k and is defined by the following formula:

$$r(k) = \pi^{-1/2} \left[\Gamma\left(1 + \frac{1}{n}\right) m(S)\sigma \frac{\ln kN}{kN} \right]^{1/n}, \qquad (4.24)$$

where S is the domain of the objective function, $m(S)$ the Lebesque measure of S, Γ the gamma function, σ a positive number, and N the sample size.

Once a set of starting points is selected, local optimization is applied. Here, it is assumed that the local optimization procedure can always find a local optimal solution x^* when starting with a point x that is within the region of attraction of x^*. Let w be the number of local minima found by the method in the kth iteration. Then, based on a Bayesian estimate, after the kth iteration, the total number of local minima found by the method and the portion of S covered by the regions of attraction of the local optimal solutions are given by

$$v = \frac{w(kN - 1)}{kN - w - 2} \quad \text{and}$$

$$u = \frac{(kN - w - 1)(kN + w)}{kN(kN - 1)}, \qquad (4.25)$$

respectively. With these estimates, a stopping rule can be given to terminate the method after the kth iteration when $v \leq 1.5w$ and $u \geq 0.995$.

Strong convergence results can be obtained for the multi-level single-linkage method. If all requirements for the selection of the starting points and the termination of the method are met, then the method can find, with probability 1, all of the isolated local minima including the global minimum of the objective function within a finite number of iterations.

4.3.3. *Branch and bound*

Branch and bound is a scheme widely used in integer and combinatorial optimization, and can also be applied to global optimization. First, a pair of upper and lower bounds must be estimated for the global minimum. Then, the domain of the objective function is divided to find better bounds. The process continues until it finds a pair of bounds that is the closest possible to the global minimum. More specifically, at each step, the domain of the objective function is divided into a set of small regions. For each of these small regions, a pair of upper and lower bounds for the global minimum within that region is estimated. If the lower bound is already higher than the currently best upper bound, the upper bound within that region cannot be better than the currently best upper bound, and the search in that region can stop. Otherwise, if the upper bound within that region is lower than the currently best upper bound, the best upper bound is set to the upper bound within that region, and the region is divided into a set of smaller regions for further estimations.

Key to the branch-and-bound scheme are the branching strategy and the bounding procedure. Branching includes dividing a region and choosing the next subregion for further estimation. In depth-first branching, the first child region is always selected for the first estimation. In breadth-first branching, the regions in the same generation are estimated first, and in the next generation second, and so on and so forth. In best-first branching, the most preferred region among all available regions, judged by some criteria such as the lower bound, the upper bound, or the difference between the lower and upper bounds, is always selected for the first estimation. The preferred branching strategy depends on the specific problem and the implementation.

The upper bound for global optimization can be obtained relatively easily by using local optimization because any local minimum provides an upper bound for the global minimum. On the other hand, the lower bound is relatively harder to obtain. Different methods have been developed. One of these is the so-called majorization method. In this method, a lower-bounding function for the objective function is constructed first, for example, by using a quadratic function interpolated with a sample set of objective function values. The minimum of the lower-bounding function can then be obtained and used as a lower bound for the global minimum.

4.3.4. *Simulated annealing*

Simulated annealing is one of the most popular methods for global optimization. Perhaps due to its physical intuitiveness, it is especially favored by scientists in scientific applications. In physical annealing, a system is first heated to a melting state and then cooled down slowly. If the process is slow enough for the system to reach equilibrium, the system will eventually settle down to a ground state with minimal energy, when the temperature is lowered to zero. A simulated annealing algorithm considers the objective function for global optimization as the energy function of an imagined physical system and mimics the physical annealing process so that, in the end, the global minimum of the objective function can be located by reaching the global energy minimum of the imagined physical system.

A typical simulated annealing algorithm works as follows. First, it repeats a loop where the temperature is decreased in every loop cycle. Within the loop, at temperature T, the algorithm mimics the cooling process of the physical system by making a sequence of random changes on the system. Each random change is made by perturbing the current system state by a random amount. Let the energy difference due to the state change be denoted as Δf. Then, the new state is accepted with a probability e equal to $\exp(-\Delta f/kT)$, where k is the Boltzmann constant and T is the temperature.

Note that in each random change, if the energy difference is negative, the function value decreases. Thus, the new state will always be accepted because the value of e is greater than or equal to one. On the other hand, if the energy difference is positive, the function value increases, but the new state is still accepted with some probability because the value of e is a fraction. The latter makes the algorithm different from a conventional local optimization algorithm, which only accepts a state that decreases the function value. In this way, the algorithm is able to search for the global minimum in both the descent and ascent directions, preventing it from being trapped in a local minimum surrounded by only ascent directions.

Let f be the objective function and x the variable. In simulated annealing, f is considered as the energy function of an imagined physical system and x as the system's state. It can be shown that by following a simulated annealing algorithm, the probability for the algorithm to reach a state x of the system at temperature T is subject to the Boltzmann distribution $p(x) = \exp(-f(x)/kT)/Z$, where Z is the normalization factor. Based on this property, it can also be shown that if the temperature is lowered gradually and the number of random changes at each temperature is made sufficiently large, the probability for the algorithm to reach the global minimum of f converges to one.

4.4. Energy Transformation

No matter how effective a global optimization algorithm is, the global minimum will still be costly to obtain if the objective function has too many local minima. Therefore, an important research subject that has been pursued in recent years is to use special techniques to change the function into one with fewer local minima so that the search for the global minimum can be more effective and efficient.

4.4.1. *Integral transform*

One of the techniques for energy transformation is to use a special integral transform called the Gaussian transform. Given an objective function f, the Gaussian transform for f is a parametrized integral of the product of the function f with a Gaussian distribution function g_λ with the standard deviation equal to λ (Fig. 4.4). Let the transformed function be denoted by $\langle f \rangle_\lambda$. Then,

$$\langle f \rangle_\lambda(x) = \int_{R^n} f(x')g_\lambda(x - x')dx', \tag{4.26}$$

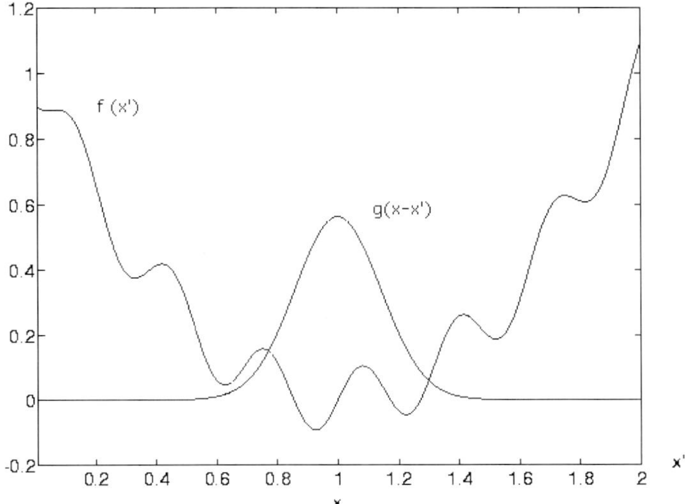

Fig. 4.4. Function transformation. The transformed function is an average of the original function sampled by a probability distribution function.

where $g_\lambda(x-x') = c_\lambda \exp(-\,||x-x'||^2/\lambda^2)$, with c_λ being the normalization constant for g_λ.

Based on the definition of the Gaussian transform, the value of the transformed function $\langle f \rangle_\lambda$ at any point x is in fact the average value of the original function f sampled by a Gaussian distribution function within a neighborhood of x. The most significant part of the sampled neighborhood corresponds to the dominant region of the Gaussian distribution function, which is determined by the value of λ. So, if λ equals zero, nothing is averaged and the transformed function is therefore exactly equal to the original function. Otherwise, the small variations of the original function are averaged out, and the transformed function should become smoother than the original function. If λ is sufficiently large, the transformed function may have fewer local minima. In this case, the global minimum may be easier to find in the transformed function. Then, by gradually changing the transformed function back to the original function, the change in the global minimum may be traced to find the global minimum of the original function.

4.4.2. *Solution to the diffusion equation*

The transformation of a function can be viewed as if the function is geometrically deformed into a smoother or flatter function as the parameter λ is changed from zero to a large positive number. It can also be considered physically as a diffusion process, with the given function as a diffusing heat function for some physical system.

Let the time parameter for the system be denoted by t. Then, the process can be described mathematically by the diffusion equation

$$u_t(x, t) = u_{xx}(x, t), \quad x \in R^n, \quad t > 0,$$
$$u(x, t) = f(x), \quad x \in R^n, \quad t = 0, \tag{4.27}$$

where u is the heat function and f is the given function. The heat function u is a function of space and time variables x and t, respectively, and the given function f is considered as the initial heat function to be diffused.

A standard solution to the above differential equation is

$$u(x, t) = \int_{R^n} K_t(x - x') f(x') \mathrm{d}x', \quad x \in R^n, \quad t \geq 0, \tag{4.28}$$

where $K_t(x - x') = c_t \exp(-||x-x'||/4t)$, with c_t being the normalization constant for K_t. If we let $4t = \lambda^2$, the solution to the diffusion equation gives exactly the Gaussian transform of the given function.

4.4.3. *Smoothing properties*

Note that the Gaussian transform $\langle f \rangle_\lambda$ on a given function f is in fact a convolution of the function f and the Gaussian distribution function g_λ. If we make a Fourier transformation on the transformed function $\langle f \rangle_\lambda$, the transform at a specific frequency ω, $FT[\langle f \rangle_\lambda](\omega)$, will be the product of the Fourier transform for the given function, $FT[f](\omega)$, and the Fourier transform for the Gaussian distribution function, $FT[g_\lambda](\omega)$, at the corresponding frequency. Note that

$$FT[g_\lambda](\omega) = \frac{1}{(2\pi)^{1/2n}} \int_{R^n} c_\lambda \exp(-||x||^2/\lambda^2) \exp(-i\omega \cdot x) dx$$

$$= \frac{1}{(2\pi)^{1/2n}} c_\lambda \exp(-\lambda^2 ||\omega||^2/4). \tag{4.29}$$

Therefore,

$$FT[\langle f \rangle_\lambda](\omega) = FT[g_\lambda](\omega) \times FT[f](\omega)$$

$$= \frac{1}{(2\pi)^{1/2n}} c_\lambda \exp(-\lambda^2 ||\omega||^2/4) \times FT[f](\omega). \tag{4.30}$$

From the above relationship, we can see that if ω is large, the Fourier transform for the transformed function $\langle f \rangle_\lambda$ will be small. On the other hand, for a fixed frequency ω, if λ is large, the Fourier transform for the transformed function $\langle f \rangle_\lambda$ will also be small. All of these properties imply that after the transformation, the high-frequency components of the original function will be reduced in the transformed function, especially when a large λ value is used. This is why the transformed function may become smoother with fewer local minima than the original function.

4.4.4. *Computation of transformation*

The Gaussian transform involves a high-dimensional integral, which cannot be computed efficiently for a dimension higher than three. In practice, a given objective function to be transformed is often defined in a space of more than several hundred dimensions, so its transformation may not be possible in general. However, a large class of partially separable functions called decomposable functions can always be transformed, even in high-dimensional space.

A decomposable function is the sum of the products of some one-dimensional (1D) functions. For such a class of functions, the Gaussian transform is in fact equal to the sum of the products of the transformations of the 1D functions. The 1D transformation can always be computed either analytically or numerically. Therefore, the transformation of decomposable functions can always be done efficiently.

Formally, if f is a decomposable function and f is equal to the sum of the products of the 1D functions $g_{i,j}$, then

$$f(x) = \sum_i \prod_j g_{i,j}(x_{i,j}), \qquad (4.31)$$

where $x_{i,j}$ is a 1D variable, meaning the jth variable in the ith product. It is easy to verify that the Gaussian transform of f is

$$\langle f \rangle_\lambda(x) = \sum_i \prod_j \langle g_{i,j} \rangle_\lambda(x_{i,j}). \qquad (4.32)$$

Apparently, all polynomial functions are decomposable functions. Therefore, all polynomial functions can be transformed.

As an example, let us consider the following Griewant function:

$$f(x_1, x_2) = (x_1^2 + x_2^2)/200 + 1 - \cos(x_1/\sqrt{1}) \cos(x_2/\sqrt{2}).$$

This function has many local minima because of the cosine part of the function. It has been widely used as a test function for global optimization. If a general global optimization algorithm is applied to the function, a large number of local minima must be examined before the global minimum is eventually found. However, if we use the Gaussian transform, the function can be transformed into the following form:

$$\langle f \rangle_\lambda(x_1, x_2) = \frac{(x_1^2 + x_2^2 + \lambda^2)/200}{1 - \exp(-3\lambda^2/8) \cos(x_1/\sqrt{1}) \cos(x_2/\sqrt{2})}.$$

If λ is equal to 0, the transformed function is the same as the original function. Otherwise, if λ is increased from 0 to 1, 1 to 2, 2 to 3, . . . , the transformed function will become smoother and smoother, and eventually become convex (Fig. 4.5). Then, with just a local optimization algorithm, the global minimum of the transformed function can be found. With this global minimum as the starting point, the local optimization algorithm can be applied again to the transformed

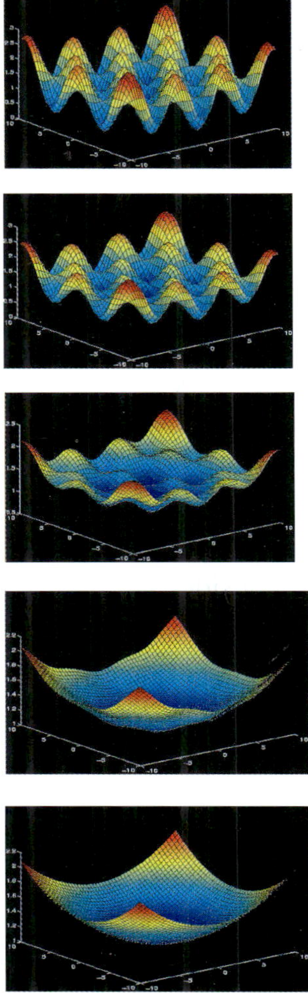

Fig. 4.5. Transformation of Griewant function. A 2D Griewant function is transformed into a convex function with increasing parameter λ from 0 to 5.

function with a slightly lowered λ value, and the global minimum of the function can again be found. The process can be repeated until the λ value is decreased to 0, when the transformed function becomes the original function, and the global minimum of the original function can be found.

Selected Further Readings

Potential energy minimization

Schlick T, *Molecular Modelling and Simulation: An Interdisciplinary Guide*, Springer, 2003.

Brooks III CL, Karplus M, Pettitt BM, *Proteins: A Theoretical Perspective of Dynamics, Structure, and Thermodynamics*, Advances in Chemical Physics, Vol. 71, Wiley, 2004.

Anfinsen CB, Studies on the principles that govern the folding of protein chains, Nobel Lecture, 1972.

Levitt M, Lifson S, Refinement of protein conformations using a macromolecular energy minimization procedure, *J Mol Biol* **46**: 269–279, 1969.

Levitt M, Protein folding by restrained energy minimization and molecular dynamics, *J Mol Biol* **170**: 723–764, 1983.

Scheraga HA, Approaches to the multiple-minima problem in conformational energy calculations on polypeptides and proteins, in *Biological and Artificial Intelligence Systems*, Clementi E, Chin S (eds.), ESCOM Science Publishers, pp. 1–14, 1988.

Wales DJ, Scheraga HA, Global optimization of clusters, crystals, and biomolecules, *Science* **285**: 1368–1372, 1999.

Wolynes PG, Folding funnels and energy landscapes of larger proteins within the capillarity approximation, *Proc Natl Acad Sci USA* **94**: 6170–6175, 1997.

Onuchic J, Wolynes, P, Theory of protein folding, *Curr Opin Struct Biol* **14**: 70–75, 2004.

Local optimization

Ortega JM, Rheinboldt WC, *Iterative Solution of Nonlinear Equations in Several Variables*, SIAM, 1987.

Dennis JE, Schnabel RB, *Numerical Methods for Unconstrained Optimization and Nonlinear Equations*, SIAM, 1996.

Fletcher R, *Practical Methods of Optimization*, Wiley, 2000.

Nocedal J, Wright S, *Numerical Optimization*, Springer, 2006.

Fletcher R, Reeves CM, Function minimization by conjugate gradients, *Computer J* **7**: 149–154, 1964.

Dennis Jr JE, Moré JJ, Quasi-Newton methods, motivation and theory, *SIAM Rev* **19**: 46–89, 1977.

Global optimization

Horst R, Pardalos PM, Nguyen T, *Introduction to Global Optimization*, Springer, 2006.

Zabinsky ZB, *Stochastic Adaptive Search for Global Optimization*, Springer, 2006.

Van Laarhoven PJM, Aarts EHL, *Simulated Annealing: Theory and Applications*, Kluwer Academic Publishers, 1987.

Rinnooy Kan AHG, Timmer GT, A stochastic approach to global optimization, in *Numerical Optimization*, Boggs P, Byrd R, Schnabel RB (eds.), SIAM, pp. 254–262, 1984.

Kirkpatrick S, Gelatt Jr CD, Vecchi MP, Optimization by simulated annealing, *Science* **220**: 671–680, 1983.

Function transformation

Piela L, Kostrowicki J, Scheraga HA, The multiple-minima problem in the conformational analysis of molecules: deformation of the potential energy hyper-surface by the diffusion equation method, *J Phys Chem* **93**: 3339–3346, 1989.

Kostrowicki J, Scheraga HA, Application of the diffusion equation method for global optimization to oligopeptides, *J Phys Chem* **96**: 7442–7449, 1992.

Kostrowicki J, Scheraga HA, Some approaches to the multiple-minima problem in protein folding, in *Global Minimization of Non-convex Energy Functions: Molecular Conformation and Protein Folding*, Pardalos PM, Shalloway D, Xue G (eds.), American Mathematical Society, pp. 123–132, 1996.

Shalloway D, Application of the renormalization group to deterministic global minimization of molecular conformation energy functions, *J Glob Optim* **2**: 281–311, 1992.

Shalloway D, Packet annealing: A deterministic method for global minimization. Application to molecular conformation, in *Recent Advances in Global Optimization*, Floudas C, Pardalos P (eds.), Princeton University Press, pp. 433–477, 1992.

Coleman T, Shalloway D, Wu Z, Isotropic effective energy simulated annealing searches for low energy molecular cluster states, *Comput Optim Appl* **2**: 145–170, 1993.

Coleman T, Shalloway D, Wu Z, A parallel buildup algorithm for global energy minimization of molecular clusters using effective energy simulated annealing, *J Glob Optim* **4**: 171–185, 1994.

Coleman T, Wu Z, Parallel continuation-based global optimization for molecular conformation and protein folding, *J Glob Optim* **8**: 49–65, 1996.

Wu Z, The effective energy transformation scheme as a special continuation approach to global optimization with application to molecular conformation, *SIAM J Optim* **6**: 748–768, 1996.

Moré J, Wu Z, Smoothing techniques for macromolecular global optimization, in *Nonlinear Optimization and Applications*, Di Pillo G, Gianessi F (eds.), Plenum Press, pp. 297–312, 1996.

Moré J, Wu Z, Global continuation for distance geometry problems, *SIAM J Optim* **7**: 814–836, 1997.

Head-Gordon T, Arrecis J, Stillinger FH, A strategy for finding classes of minima on a hyper-surface: Implications for approaches to the protein folding problem, *Proc Natl Acad Sci USA* **88**: 11076–11080, 1991.

Head-Gordon H, Stillinger FH, Predicting polypeptide and protein structures from amino acid sequences: Ant-lion method applied to melittin, *Biopolymers* **33**: 293–303, 1993.

Straub J, Ma J, Amara P, Simulated annealing using coarse-grained classical dynamics: Fokker–Planck and Smoluchowski dynamics in the Gaussian density approximation, *J Chem Phys* **103**: 1574–1581, 1995.

Straub JE, Optimization techniques with applications to proteins, in *New Developments in Theoretical Studies of Proteins*, Elber R (ed.), World Scientific Publishing Co., pp. 137–196, 1996.

Shao CS, Byrd RH, Eskow E, Schnabel RB, Global optimization for molecular clusters using a new smoothing approach, in *Large Scale Optimization with Applications*, Biegler L, Coleman T, Con A, Santosa F (eds.), Springer-Verlag, pp. 163–199, 1997.

Azmi A, Byrd R, Eskow E, Schnabel R, Crivelli S, Philip T, Head-Gordon T, Predicting protein tertiary structure using a global optimization algorithm with smoothing, in *Optimization in Computational Chemistry and Molecular Biology: Local and Global Approaches*, Floudas C, Pardalos P (eds.), Kluwer Academic Publishers, pp. 1–18, 2000.

Chapter 5

Molecular Dynamics Simulation

The dynamic behavior of a protein is as essential as the structure for the study of its function. A protein changes from an arbitrary state to an equilibrium state, by folding, and also fluctuates around an equilibrium state. The former type of dynamics is called general dynamics, and the latter thermodynamics. The dynamic properties of proteins can be studied through molecular dynamics simulation.

5.1. Equations of Motion

The simulation of molecular dynamics is based on an atomic-level classical mechanical model for the molecule. Let $x(t)$ be the configuration of the molecule at time t, $x = \{x_i : x_i = (x_{i,1}, x_{i,2}, x_{i,3})^{\mathrm{T}}, i = 1, 2, \ldots, n\}$, where x_i is the position vector of atom i and n is the total number of atoms in the molecule. Then, the molecular motion can be described as a collection of movements of the atoms in the molecule, as given in the following equations:

$$m_i x_i'' = f_i(x_1, x_2, \ldots, x_n),$$
$$f_i(x_1, x_2, \ldots, x_n) = -\partial E(x_1, x_2, \ldots, x_n)/\partial x_i, \quad i = 1, 2, \ldots, n, \tag{5.1}$$

where m_i and f_i are the mass and force of atom i, respectively, and E is the potential energy.

125

The solution to Eq. (5.1) is in general not unique, unless additional conditions are given. Two types of conditions are often imposed on the solution: conditions on the initial positions and velocities of the atoms, or conditions on the initial and ending positions of the atoms. The former are called initial conditions, and the latter are called boundary conditions. The problem of solving Eq. (5.1) with a set of initial conditions is called an initial-value problem, and the problem with a set of boundary conditions is called a boundary-value problem.

5.1.1. *Least-action principle*

Based on the theory of classical mechanics, the trajectory of molecular motion between two molecular states minimizes the total action of the molecule. More specifically, let $x(t)$ be the configuration of the molecule at time t, $x = \{x_i : x_i = (x_{i,1}, x_{i,2}, x_{i,3})^T, i = 1, 2, \ldots, n\}$. Given the beginning and ending times t_0 and t_e, respectively, $x(t)$ in $[t_0, t_e]$ defines a molecular trajectory connecting the two molecular states $x^0 = x(t_0)$ and $x^e = x(t_e)$. Let $L(x, x', t)$ be the difference between the kinetic and potential energies of the molecule at time t. The functional L is called the Lagrangian of the molecule. Let S be the action of the molecule in $[t_0, t_e]$. Then, S is defined as the integral of the Lagrangian in $[t_0, t_e]$. According to the least-action principle, the trajectory x minimizes the action S of the system along x, and therefore x is a solution to the so-called least-action problem:

$$\min S(x) = \int_{t_0}^{t_e} L(x, x', t) \, dt. \tag{5.2}$$

5.1.2. *Principle of variation*

Let L be a continuously differentiable functional. Let x be a solution to the least-action problem. Let δx be a small variation of x, and $\delta x(t_0) = \delta x(t_e) = 0$. Based on the principle of variation (Fig. 5.1), the necessary condition for x to be a solution of the least-action problem is that

$$\delta S = \int_{t_0}^{t_e} \left[\frac{\partial L(x, x', t)}{\partial x} \delta x + \frac{\partial L(x, x', t)}{\partial x'} \delta x' \right] dt = 0. \tag{5.3}$$

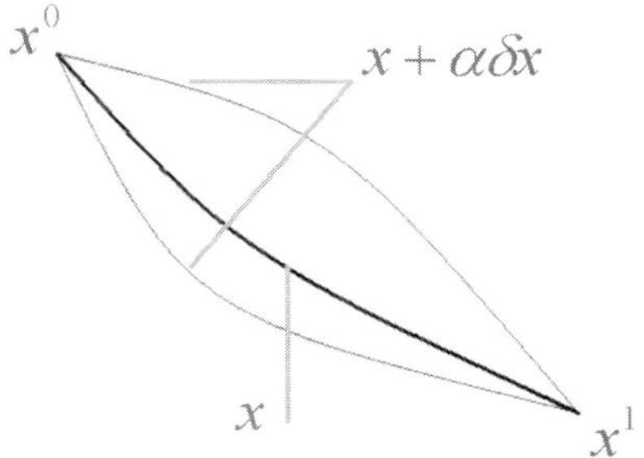

Fig. 5.1. Principle of variation. A small variation of the trajectory results in a small variation in the action.

Since $\delta x' = \delta\,dx/dt = d\delta x/dt$, we obtain, after integrating the second term of the integrand by parts,

$$\delta S = \int_{t_0}^{t_e} \left(\frac{\partial L(x, x', t)}{\partial x} - \frac{d}{dt}\left[\frac{\partial L(x, x', t)}{\partial x'} \right] \right) \delta x \, dt = 0. \tag{5.4}$$

Since δS should be zero for all δx, the integrand must be zero, and it follows that

$$\frac{\partial L(x, x', t)}{\partial x} - \frac{d}{dt}\left[\frac{\partial L(x, x', t)}{\partial x'} \right] = 0. \tag{5.5}$$

The above equation is the famous Euler–Lagrange equation in classical mechanics.

5.1.3. *Equation for molecular motion*

Let $L = x'^{\mathrm{T}} M x'/2 - E(x)$, where M is the mass matrix of the molecule and E is the potential energy. Based on the results in the previous section and on the facts that $d(\partial L/\partial x')/dt = Mx''$ and $\partial L/\partial x = -\nabla E$, then x necessarily satisfies the following equation of motion:

$$Mx'' = -\nabla E(x), \tag{5.6}$$

or equivalently,

$$m_i x_i'' = f_i(x_1, x_2, \ldots, x_n), \quad f_i = -\partial E/\partial x_i, \quad i = 1, 2, \ldots, n, \tag{5.7}$$

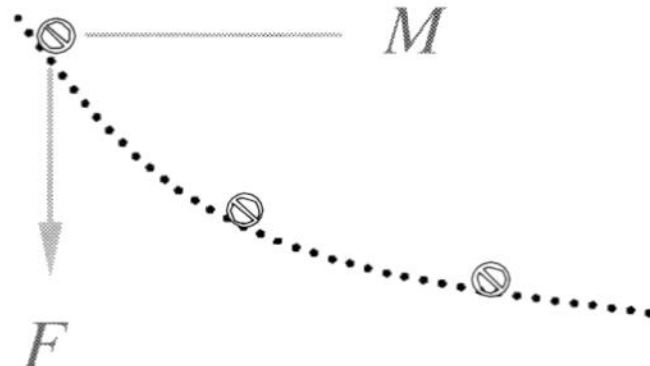

Fig. 5.2. Newton's law of motion. The mass M multiplied by the acceleration of the mass should be equal to the force F.

where, again, m_i and f_i are the mass and force for atom i, respectively, and $M = \text{diag}[m_1, m_2, \ldots, m_n]$ (Fig. 5.2).

Note that the above results imply that a molecular trajectory which minimizes the molecular action between two molecular states necessarily satisfies the classical mechanical equation of motion. In other words, the solution of the equation of motion can be considered as an attempt to obtain a molecular trajectory that minimizes the action of the molecule on the entire trajectory.

5.1.4. *Force field calculation*

To compute the forces on the atoms, we need to differentiate the energy function. Let $f = \{f_i : f_i = (f_{i,1}, f_{i,2}, f_{i,3})^{\mathrm{T}}, i = 1, 2, \ldots, n\}$. Then, f is equal to the negative gradient of the energy function E. The computation of the energy function generally requires an order of n^2 floating-point operations because it has an order of n^2 pairwise interaction terms such as the electrostatic and van der Waals potential terms.

Without loss of generality, let us assume that the energy function has the following form:

$$E = \sum_{i=1, j>i}^{n} h_{i,j}(\|x_i - x_j\|), \tag{5.8}$$

with an order of n^2 interaction terms dependent on the distances between pairs of atoms in the molecule. Then, one energy function evaluation requires an order of n^2 floating-point operations.

Computing the gradient of a function usually requires n times the order of operations required for one function evaluation, but because the energy function E is partially separable, the computation of the gradient can actually be arranged to require the same number of floating-point operations as for one energy function evaluation (on the order of n^2 floating-point operations). In fact, the partial derivatives of E have the following form:

$$\frac{\partial E}{\partial x_k} = \sum_{j=k+1}^{n} \frac{h'_{j,k}(||x_k - x_j||)x_k}{||x_k - x_j||} - \sum_{i=1}^{k} \frac{h'_{i,k}(||x_i - x_k||)x_k}{||x_i - x_k||},$$

$$k = 1, 2, \ldots, n. \qquad (5.9)$$

Therefore, each gradient computation requires only an order of n floating-point operations.

5.2. Initial-Value Problem

The dynamic properties of a protein can be studied by forming an initial-value problem with the equation of motion. Typically, the problem is defined as

$$m_i x_i'' = f_i(x_1, x_2, \ldots, x_n),$$

$$x_i(t_0) = x_i^0, \quad v_i(t_0) = v_i^0, \quad i = 1, 2, \ldots, n, \qquad (5.10)$$

where x_i and v_i are the positions and velocities of the atoms, respectively, and $v_i = dx_i/dt$; and m_i and f_i are the masses and forces, respectively, and $f_i = -\partial E/\partial x_i$. Usually, the equation of motion has infinitely many solutions. However, with a set of initial conditions as given, a unique solution can usually be obtained.

5.2.1. *Initial positions and velocities*

The conditions for the initial positions can be obtained by assigning an appropriate set of coordinates for the atoms, for example, the coordinates of the atoms determined from X-ray crystallography or nuclear magnetic resonance (NMR) experiments. The conditions for the initial velocities depend on the temperature, for example, a room temperature or a physiological temperature. The temperature is correlated with the average velocities of the atoms.

Let T be the temperature. Then, the average kinetic energy of the molecule should be $3nk_BT/2$, where k_B is the Boltzmann constant and n is the number of atoms in the molecule. Let v_i be the velocity vector of atom i. The kinetic energy of atom i is $m_i\|v_i\|^2/2$, and the total kinetic energy of the molecule is

$$E_k = \frac{\sum_{i=1}^{n} m_i\|v_i\|^2}{2}. \tag{5.11}$$

In order to generate a set of initial velocities so that the expected kinetic energy equals $3nk_BT/2$, we can first generate each v_i so that v_i has a normal distribution $N(\mu, \sigma)$ with mean $\mu = 0$ and variance $\sigma^2 = 3k_BT/m_i$, i.e. $\langle\|v_i\|^2\rangle = 3k_BT/m_i$. Then, by the central limit theorem, the expected kinetic energy of the molecule will equal the sum of the expected values of $m_i\|v_i\|^2/2$. We then have

$$\langle E_k \rangle = \frac{\sum_{i=1}^{n} \langle m_i\|v_i\|^2\rangle}{2} = \frac{3nk_BT}{2}. \tag{5.12}$$

5.2.2. *The Verlet algorithm*

Verlet (1967) developed an algorithm, now called the Verlet algorithm, for numerically integrating the equation of motion in problem (5.10), starting with the initial positions and velocities for the atoms. There are two versions of this algorithm, the position Verlet and velocity Verlet, as given by the following formulas:

$$x_i^{k+1} = 2x_i^k - x_i^{k-1} + \frac{h^2 f_i^k}{m_i},$$

$$\text{Position Verlet} \qquad v_i^{k+1} = v_i^k + \frac{h f_i^k}{m_i}, \tag{5.13}$$

$$i = 1, 2, \ldots, n, \quad k = 1, 2, \ldots$$

$$x_i^{k+1} = x_i^k + h v_i^k + \frac{h^2 f_i^k}{2 m_i},$$

Velocity Verlet $\qquad v_i^{k+1} = v_i^k + \frac{h(f_i^k + f_i^{k+1})}{2 m_i},$ $\qquad$ (5.14)

$$i = 1, 2, \ldots, n, \quad k = 0, 1, \ldots$$

where h is the step size, $x_i^k = x_i(t_k)$, $v_i^k = v_i(t_k)$, and $f_i^k = f_i(x_1^k, x_2^k, \ldots, x_n^k)$, with $t_k = t_0 + kh$ (Fig. 5.3). If we put the formulas into vector forms, we have

$$x^{k+1} = 2x^k - x^{k-1} + h^2 M^{-1} f^k,$$

Position Verlet $\qquad v^{k+1} = v^k + h M^{-1} f^k,$ $\qquad$ (5.15)

$$k = 1, 2, \ldots$$

$$x^{k+1} = x^k + h v^k + \frac{h^2 M^{-1} f^k}{2},$$

Velocity Verlet $\qquad v^{k+1} = v^k + \frac{h M^{-1}(f^k + f^{k+1})}{2},$ $\qquad$ (5.16)

$$k = 1, 2, \ldots$$

where $x = \{x_i, i = 1, 2, \ldots, n\}$, $v = \{v_i, i = 1, 2, \ldots, n\}$, $f = \{f_i, i = 1, 2, \ldots, n\}$, $M = \text{diag}[m_1, m_2, \ldots, m_n]$, $x^k = x(t_k)$, $v^k = v(t_k)$, and $f^k = f(x^k)$.

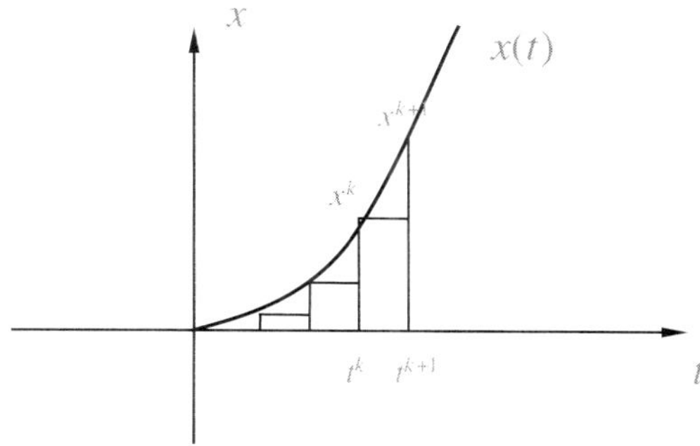

Fig. 5.3. Numerical method. The trajectory is discretized over a set of time intervals, where the points on the trajectory can be approximately calculated.

We now examine the accuracy of the Verlet formulas. We first consider the position Verlet formula. Based on the Taylor theory,

$$x(t_k + h) = x(t_k) + hx'(t_k) + \frac{h^2 x''(t_k)}{2} + \frac{h^3 x^{(3)}(t_k)}{6} + \Delta(h^4) \quad (5.17)$$

and

$$x(t_k - h) = x(t_k) - hx'(t_k) + \frac{h^2 x''(t_k)}{2} - \frac{h^3 x^{(3)}(t_k)}{6} + \Delta(h^4). \quad (5.18)$$

By adding the two formulas, we obtain

$$x(t_k + h) - x(t_k - h) = 2x(t_k) + h^2 x''(t_k) + \Delta(h^4), \quad (5.19)$$

which is equivalent to

$$x^{k+1} = 2x^k - x^{k-1} + h^2 M^{-1} f^k + \Delta(h^4). \quad (5.20)$$

Without the $\Delta(h^4)$ term, the formula is the same as the position update in the position Verlet formula. Therefore, the position Verlet has a fourth-order accuracy for the calculation of the positions. However, the Taylor expansion for the velocity at t_k is

$$v(t_k + h) = v(t_k) + hv'(t_k) + \Delta(h^2), \quad (5.21)$$

which is equivalent to

$$v^{k+1} = v^k + hM^{-1} f^k + \Delta(h^2). \quad (5.22)$$

Again, without the $\Delta(h^2)$ term, the formula is the same as the velocity update in the position Verlet formula. Therefore, the position Verlet has a second-order accuracy for the calculation of the velocities.

We now consider the velocity Verlet formula. It is easy to see that the position update in the velocity Verlet formula is a third-order approximation, because it contains terms up to the second order in the Taylor expansion of the position:

$$x(t_k + h) = x(t_k) + hx'(t_k) + \frac{h^2 x''(t_k)}{2} + \Delta(h^3), \quad (5.23)$$

which is equivalent to

$$x^{k+1} = x^k + hv^k + \frac{h^2 M^{-1} f^k}{2} + \Delta(h^3). \quad (5.24)$$

For the velocity update, the Taylor expansions of the velocity at t_k and $t_k - h$ are

$$v(t_k + h) = v(t_k) + hv'(t_k) + \frac{h^2 v''(t_k)}{2} + \Delta(h^3) \qquad (5.25)$$

and

$$v(t_k) = v(t_k + h) - hv'(t_k + h) + \frac{h^2 v''(t_k + h)}{2} + \Delta(h^3). \qquad (5.26)$$

By subtracting the second formula from the first, we obtain

$$v(t_k + h) = v(t_k) + \frac{h[v'(t_k) + v'(t_k + h)]}{2} + \Delta(h^3), \qquad (5.27)$$

or equivalently,

$$v^{k+1} = v^k + \frac{hM^{-1}[f^k + f^{k+1}]}{2} + \Delta(h^3). \qquad (5.28)$$

Without the $\Delta(h^3)$ term, the formula is the same as the velocity update in the velocity Verlet formula. Therefore, the velocity Verlet has a third-order accuracy for the calculation of the velocities.

Although the position Verlet has a higher accuracy for the calculation of the positions, the velocity Verlet has a higher accuracy for the calculation of the velocities while not losing much accuracy for the positions. The latter also has more favorable properties for preserving the energy and volume of the system, and is therefore preferred in practice.

In any case, to maintain the accuracy of the calculation, the step size in the Verlet updates has to be small, on the order of femtosecond (1.0×10^{-15} second). This limits the application of molecular dynamics simulation for protein motions with long time frames. For example, a nanosecond simulation would require about one million steps of calculation, while many protein motions such as folding require more than milliseconds to complete.

On the other hand, at every step of the Verlet algorithm, the forces on the atoms have to be calculated at least once. Each force field evaluation requires an order of n^2 floating-point operations if no approximation methods are used. For proteins, n is on the order of 10 000, and therefore one force field evaluation may require an order of 100 million floating-point operations.

5.2.3. *Leap-frog algorithm*

The leap-frog algorithm is almost the same as the Verlet algorithm, except that the points chosen for evaluating the positions and the velocities cross-over with each other. Let $k - \frac{1}{2}$ or $k + \frac{1}{2}$ be indices for the middle points between points $k - 1$ and k, and between points k and $k + 1$. Then, the leap-frog versions of the position and velocity Verlet algorithms have the following update formulas:

Position leap-frog
$$
\begin{aligned}
x^{k+\frac{1}{2}} &= x^k + \frac{hv^k}{2}, \\
v^{k+1} &= v^k + hM^{-1}f^{k+\frac{1}{2}}, \\
x^{k+1} &= x^{k+\frac{1}{2}} + \frac{hv^{k+1}}{2}.
\end{aligned}
\tag{5.29}
$$

Velocity leap-frog
$$
\begin{aligned}
v^{k+\frac{1}{2}} &= v^k + hM^{-1}f^k, \\
x^{k+1} &= x^k + hv^{k+\frac{1}{2}}, \\
v^{k+1} &= v^{k+\frac{1}{2}} + \frac{hM^{-1}f^{k+1}}{2}.
\end{aligned}
\tag{5.30}
$$

The above formulas can also be reorganized into the following simpler forms:

Position leap-frog
$$
\begin{aligned}
x^{k+1} &= x^k + hv^k + \frac{h^2 M^{-1} f^{k+\frac{1}{2}}}{2}, \\
v^{k+1} &= v^k + hM^{-1}f^{k+\frac{1}{2}}.
\end{aligned}
\tag{5.31}
$$

Velocity leap-frog
$$
\begin{aligned}
v^{k+\frac{1}{2}} &= v^{k-\frac{1}{2}} + hM^{-1}f^k, \\
x^{k+1} &= x^k + hv^{k+\frac{1}{2}}.
\end{aligned}
\tag{5.32}
$$

The advantage of using the leap-frog algorithms is that they maintain better numerical properties.

5.2.4. *Shake and Rattle*

A Verlet algorithm such as the velocity Verlet preserves the energy and volume of the molecule and exhibits reasonable numerical stability for the simulation of relatively long time frames. However, the simulation has to be carried out with a small time step (in the order of 1.0×10^{-15} second) to keep up with the rapid atomic-level movements. The potential of simulating molecular motions over longer time scales beyond nanoseconds has therefore been limited.

For proteins, the bonding forces are believed to be among those responsible for the fast atomic vibrations that require small time steps to integrate. Therefore, one of the approaches to increase the step size and hence the simulation speed is to remove the bonding forces from the force field while enforcing them through a set of bond length constraints. The simulation can then be done by integrating the constrained equation of motion with larger time steps.

Let $g = \{g_j : j = 1, 2, \ldots, m\}$ be a vector of functions that can be used to define the constraints on a molecule. Let $x(t)$ be the configuration of the molecule at time t, $x = \{x_i : x_i = (x_{i,1}, x_{i,2}, x_{i,3})^{\mathrm{T}}, i = 1, 2, \ldots, n\}$, where x_i is the position vector of atom i and n is the total number of atoms in the molecule. The constrained simulation problem can then be formulated as a constrained least-action problem:

$$\min \ S(x) = \int_{t_0}^{t_e} L(x, x', t) \, dt \quad \text{subject to} \quad g(x) = 0, \qquad (5.33)$$

where S defines the action and L is the Lagrangian. By the theory of constrained optimization, a necessary condition for a molecular trajectory x between x_0 and x_e to be a solution of the constrained least-action problem is that

$$\delta S(x) - \sum_{j=1}^{m} \lambda_j \delta g_j(x) = 0, \quad g(x) = 0, \qquad (5.34)$$

where $\lambda = \{\lambda_i : i = 1, 2, \ldots, m\}$ is a vector of Lagrange multipliers.

The first equation in formula (5.34) translates further into the following extended form of the Euler–Lagrange equation:

$$\frac{\partial L(x, x', t)}{\partial x} - \frac{d}{dt}\left[\frac{\partial L(x, x', t)}{\partial x'}\right] - \sum_{j=1}^{m} \lambda_j \nabla g_j(x) = 0. \tag{5.35}$$

Let $L = x'^{\mathrm{T}} M x'/2 - E(x)$, where M is the mass matrix of the molecule and E is the potential energy. Then, the necessary condition becomes

$$M x'' = -\nabla E(x) - \sum_{j=1}^{m} \lambda_j \nabla g_j(x), \quad g(x) = 0, \tag{5.36}$$

or in another form,

$$M x'' = -\nabla E(x) - G(x)^{\mathrm{T}} \lambda, \quad g(x) = 0, \tag{5.37}$$

where $G(x)$ is the Jacobian of $g(x)$, $G = [\nabla g_1, \nabla g_2, \ldots, \nabla g_m]^{\mathrm{T}}$. For each atom, the equation can be written individually as

$$m_i x_i'' = f_i(x_1, x_2, \ldots, x_n) + \sum_{j=1}^{m} \lambda_j g_{ji}(x_1, x_2, \ldots, x_n),$$

$$g_j(x_1, x_2, \ldots, x_n) = 0, \quad j = 1, 2, \ldots, m, \quad i = 1, 2, \ldots, n, \tag{5.38}$$

where

$$f_i = -\frac{\partial E}{\partial x_i}, \quad g_{ji} = -\frac{\partial g_j}{\partial x_i}, \quad j = 1, 2, \ldots, m, \quad i = 1, 2, \ldots, n.$$

Note that in formula (5.38), the right-hand side of the first equation can be treated as a single force function (with the original force function plus a linear combination of the derivatives of the constraint functions). Therefore, the equation can be integrated in the same way as an unconstrained equation by the Verlet algorithm, except that at every step, the Lagrange multipliers λ_j, $j = 1, 2, \ldots, m$, have to be determined so that the new positions x_i, $i = 1, 2, \ldots, n$, for the atoms must satisfy the constraints $g_j(x_1, x_2, \ldots, x_n) = 0$ and $j = 1, 2, \ldots, m$.

Indeed, several algorithms have been developed along this line, including the so-called Shake and Rattle algorithms, which correspond to the position and velocity Verlet algorithms for unconstrained simulation, respectively.

Shake

$$
x_i^{k+1} = 2x_i^k - x_i^{k-1} + \frac{h^2\left(f_i^k + \sum_{j=1'}^m \lambda_j^k g_{ji}^k\right)}{m_i},
$$

$$
v_i^{k+1} = v_i^k + \frac{h\left(f_i^k + \sum_{j=1}^m \lambda_j^k g_{ji}^k\right)}{m_i},
$$

$$(5.39)$$

$$
g_j(x_1^{k+1}, x_2^{k+1}, \ldots, x_n^{k+1}) = 0, \quad j = 1, 2, \ldots, m, \quad i = 1, 2, \ldots, n.
$$

Rattle

$$
x_i^{k+1} = x_i^k + h v_i^k + \frac{h^2\left(f_i^k + \sum_{j=1}^m \lambda_j^k g_{ji}^k\right)}{2m_i},
$$

$$
v_i^{k+1} = v_i^k + \frac{h\left[f_i^k + f_i^{k+1} + \sum_{j=1}^m \left(\lambda_j^k g_{ji}^k + \lambda_j^{k+1} g_{ji}^{k+1}\right)\right]}{2m_i},
$$

$$(5.40)$$

$$
g_j(x_1^{k+1}, x_2^{k+1}, \ldots, x_n^{k+1}) = 0, \quad j = 1, 2, \ldots, m, \quad i = 1, 2, \ldots, n.
$$

In vector form, the formulas can be written as follows.

Shake

$$
x^{k+1} = 2x^k - x^{k-1} + h^2 M^{-1}\left(f^k + \sum_{j=1'}^m \lambda_j^k \nabla g_j^k\right),
$$

$$
v^{k+1} = v^k + h M^{-1}\left(f^k + \sum_{j=1'}^m \lambda_j^k \nabla g_j^k\right),
$$

$$(5.41)$$

$$
g_j(x^{k+1}) = 0, \quad j = 1, 2, \ldots, m.
$$

Rattle

$$x^{k+1} = x^k + hv^k + \frac{h^2 M^{-1}\left(f^k + \sum_{j=1}^{m} \lambda_j^k \nabla g_j^k\right)}{2},$$

(5.42)

$$v^{k+1} = v^k + \frac{hM^{-1}\left[f^k + f^{k+1} + \sum_{j=1}^{m}\left(\lambda_j^k \nabla g_j^k + \lambda_j^{k+1} \nabla g_j^{k+1}\right)\right]}{2},$$

$$g_j(x^{k+1}) = 0, \quad j = 1, 2, \ldots, m,$$

In principle, the Shake and Rattle algorithms can be used for any type of constrained dynamics simulation. In other words, the constraints can be of any type and be imposed on any part of the molecule. However, for proteins, they have been employed particularly to control fast atomic vibrations by imposing a set of constraints on the bond lengths. The time step for protein simulation can then be increased severalfold.

5.3. Boundary-Value Problem

Some protein motions such as transitions in protein conformation, protein misfolding, etc. can be simulated by forming a boundary-value problem with the equation of motion. The boundary conditions are the initial and ending positions of the atoms in the protein. A solution to the boundary-value problem is a trajectory along which the protein moves from one conformation to another. In contrast to the initial-value problem, the solution to a boundary-value problem, depending on the given conditions, may not exist; and even if it does, it may not necessarily be unique.

5.3.1. *Initial and ending positions*

Given the initial and ending positions for the atoms or, in other words, the initial and ending conformations of a protein,

$$x^0 = \{x_i^0 = (x_{i,1}^0, x_{i,2}^0, x_{i,3}^0)^{\mathrm{T}}, \ i = 1, 2, \ldots, n\},$$

(5.43)

$$x^e = \{x_i^e = (x_{i,1}^e, x_{i,2}^e, x_{i,3}^e)^{\mathrm{T}}, \ i = 1, 2, \ldots, n\},$$

a boundary-value problem for determining a molecular trajectory for the equation of motion that connects the initial and ending conformations of the protein can be expressed formally as

$$m_i x_i'' = f_i(x_1, x_2, \ldots, x_n),$$
$$x_i(t_0) = x_i^0, \quad x_i(t_e) = x_i^e, \quad i = 1, 2, \ldots, n, \tag{5.44}$$

where t_0 and t_e are the beginning and ending times, respectively. Usually, t_0 can be set to 0, and t_e needs to be determined. If a physical experiment has been conducted, t_e can be measured from the experiment. Otherwise, it can be taken as a variable to be determined in the solution. In the latter case, additional condition on the solution may be necessary; for example, t_e needs to be selected so that the total energy along the solution trajectory can be minimized.

Due to the nonlinearity of the equation of motion for a protein, it is hard to verify, given a set of boundary conditions, whether or not a solution to the boundary-value problem exists; and if it does, whether or not it is unique. The problem is therefore ill-posed and may suffer from severe numerical instability when solved by a computer with finite precision, and the solution will also be sensitive to errors in the given conditions.

5.3.2. *Finite difference*

A conventional approach to boundary-value problems is to discretize the time interval into a finite set of subintervals and make a finite difference approximation to the equation at each discretized point. Let the time interval $[t_0, t_e]$ be divided uniformly into N subintervals $[t_0, t_1], [t_1, t_2], \ldots, [t_{N-1}, t_N]$, with $t_N = t_e$. Let $x_i^k = x_i(t_k), i = 1, 2, \ldots, n, k = 0, 1, \ldots, N$. Then, a finite difference approximation to each of the equations in system (5.44) can be made as

$$\frac{x_i^{k+1} - 2x_i^k + x_i^{k-1}}{h^2} = \frac{f_i^k}{m_i}, \quad i = 1, 2, \ldots, n, \quad 1 \le k < N, \tag{5.45}$$

where h is the length of the subintervals. The approximations form a system of algebraic equations for the function values $x_i^k = x_i(t_k)$ at the discretized points t_k. The solution to this system can be used as an approximation to the solution to the original system of differential equations. However, since f_i^k are nonlinear functions of $x_1^k, x_2^k, \ldots, x_n^k$, the equations are nonlinear, and a nonlinear solver must be used for the solution of the system.

If we put Eq. (5.45) in vector form, we can obtain a more compact system:

$$\frac{x^{k+1} - 2x^k + x^{k-1}}{h^2} = M^{-1} f^k, \quad 1 \le k < N, \tag{5.46}$$

where $x = \{x_i, i = 1, 2, \ldots, n\}$, $f = \{f_i, i = 1, 2, \ldots, n\}$, $M = \text{diag}[m_1, m_2, \ldots, m_n]$, $x^k = x(t_k)$, and $f^k = f(x^k)$.

By solving Eq. (5.46) for x^{k+1}, we can obtain a formula similar to the position update in the position Verlet formula. Therefore, the equation is in a certain sense accurate for the position to the fourth order. However, for the second derivative, i.e. for the entire left-hand side of the equation, the accuracy is second order. To see this, let

$$\Delta^2 x(t_k) = \frac{x^{k+1} - 2x^k + x^{k-1}}{h^2}. \tag{5.47}$$

We can then estimate the error $\tau(t_k) = \Delta^2 x(t_k) - x''(t_k)$ by following a similar argument as for the proof of the fourth-order accuracy of the position update of the position Verlet formula. Based on that proof,

$$x(t_k + h) - x(t_k - h) = 2x(t_k) + h^2 x''(t_k) + \Delta(h^4). \tag{5.48}$$

It follows immediately that

$$\frac{x^{k+1} - 2x^k + x^{k-1}}{h^2} = x''(t_k) + \Delta(h^2), \tag{5.49}$$

and $\tau(t_k) = \Delta^2 x(t_k) - x''(t_k) = \Delta(h^2)$.

5.3.3. *Stochastic path following*

The approximation to the equation of motion by the finite difference method requires a small time step to preserve the accuracy. As a result, there can be many intermediate points and hence many variables for even a short trajectory. The stochastic path-following approach attempts to make large time steps while using optimization methods to control the approximation errors.

Let the time interval $[t_0, t_e]$ be divided uniformly into N subintervals $[t_0, t_1], [t_1, t_2], \ldots, [t_{N-1}, t_N]$, with $t_N = t_e$. Let $x_i^k = x_i(t_k)$, $i = 1, 2, \ldots, n$, $k = 0, 1, \ldots, N$. Then, $x_i^k = x_i(t_k)$ can be determined via the solution of the following least squares problem:

$$\min \sum_{k=1}^{N-1} \sum_{i=1}^{n} \left[\frac{x_i^{k+1} - 2x_i^k + x_i^{k-1}}{h^2} - \frac{f_i^k}{m_i} \right]^2, \qquad (5.50)$$

where h is the length of the subintervals or the step size.

The idea of stochastic path following is not simply to reformulate the finite difference equation (5.49) as a least squares problem (5.50). In fact, the finite difference formula is only used to make a coarse approximation to the equation, and the step size can actually be arbitrarily large, depending on how coarse the approximation is intended to be. The resulting error from such an approximation is assumed to be stochastically subject to a Gaussian distribution, and can therefore be minimized in a least squares form.

The advantage of using the stochastic path-following method is the ability to skip the detailed dynamic behaviors while tracking the motions only at a certain coarse level. By doing so, the method can be applied to molecular motions across much longer time scales than the conventional finite difference method. In fact, proteins do have motions on different time scales. For large protein motions, it makes sense to focus on motions on a large time scale. The stochastic path-following method has proven to be a scheme that can be used for such a purpose.

In any case, the variables in the least squares formulation of the stochastic path-following method include not only the coordinate vectors of the atoms, but the vectors at all of the intermediate points on the solution trajectory. Therefore, the total number of variables is $N-1$ times more than the number of variables at one molecular point. The least squares problem is also a global optimization problem and can in general be difficult to solve.

5.3.4. *Multiple shooting*

The idea of shooting is to find the molecular trajectory between two molecular states by mimicking the process of shooting a basketball from a given position to a target position by choosing the correct initial speed and direction for the ball. Let $x(t)$ be the position of a basketball at time t, m the mass, and $f(x)$ the force at position x. Then, x can be obtained as the solution to a boundary-value problem,

$$mx'' = f(x), \quad x(t_0) = x^0, \quad x(t_e) = x^e, \qquad (5.51)$$

where t_0 and t_e are the beginning and ending times, respectively; and x^0 and x^e are the beginning and ending positions, respectively. In order to find a solution to problem (5.51), first, let x be the solution to the following initial-value problem:

$$mx'' = f(x), \quad x(t_0) = x^0, \quad v(t_0) = v^0. \qquad (5.52)$$

Then, $x(t)$ depends on the initial velocity v^0, and can be written as $x(t) = x(t; v^0)$ and considered as a function of v^0. For an arbitrary v^0, $x(t)$ may not necessarily satisfy the ending condition, $x(t_e) = x^e$ in problem (5.51). In order to find such a solution or, in other words, to find a solution to the original boundary-value problem, we need to find an appropriate v^0 so that $x(t_e) = x(t_e; v^0) = x^e$. Let $\phi(v^0) = x(t_e) - x^e = x(t_e; v^0) - x^e$. Then, the problem becomes finding v^0 so that $\phi(v^0) = 0$.

In general, let x be a vector, M be a mass matrix, and f be the force field. We then have a nonlinear system of equations

$$\varphi(v^0) = x(t_e; v^0) - x^e = 0, \qquad (5.53)$$

where v^0 and x^e are all vectors, and

$$Mx'' = f(x), \quad x(t_0) = x^0, \quad v(t_0) = v^0. \qquad (5.54)$$

By solving Eq. (5.53), we can find the solution to a general boundary-value problem:

$$Mx'' = f(x), \quad x(t_0) = x^0, \quad v(t_0) = x^e. \qquad (5.55)$$

Such a method is called the single-shooting method (Fig. 5.4).

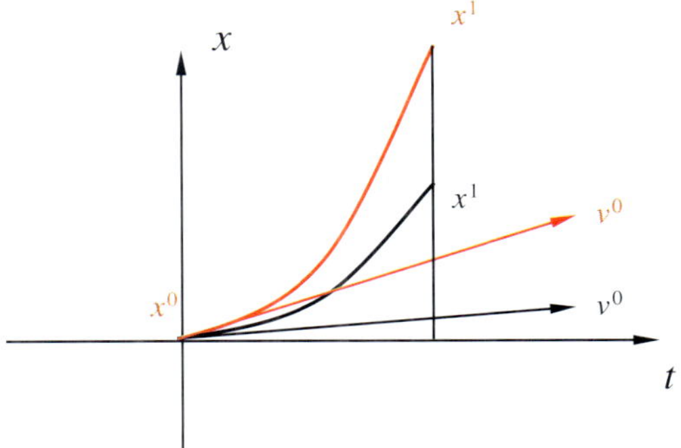

Fig. 5.4. Single shooting. The initial velocity is adjusted (or determined) so that the end position of the trajectory is equal to the given position.

The single-shooting method is not stable. We can imagine that if the time is long or, in other words, if the distance is long, the "shot" will be very sensitive to changes in the initial velocity and will have difficulty reaching the target accurately. To overcome this difficulty, we can divide the time interval into smaller subintervals, and make the shootings separately within the subintervals. Then, the positions and velocities of the solution trajectory at the intermediate points are of course all unknown. We need to determine them so that the solution trajectories obtained in the subintervals can be connected into a smooth trajectory over the entire interval. The method for obtaining such a trajectory is called the multiple-shooting method.

For a general description of the multiple-shooting method, we write the problem in the following form, without the mass matrix:

$$x'' = f(x), \quad x(t_0) = x^0, \quad x(t_e) = x^e. \tag{5.56}$$

We first divide the time interval $[t_0, t_e]$ uniformly into N subintervals $[t_0, t_1], [t_1, t_2], \ldots, [t_{N-1}, t_N]$, with $t_N = t_e$. Then, on each subinterval $[t_i, t_{i+1}]$, $0 \leq i < N$, we solve an initial-value problem:

$$x'' = f(x), \quad x(t_i) = r^{(i)}, \quad v(t_i) = s^{(i)}, \quad t_i \leq t \leq t_{i+1}. \tag{5.57}$$

Let the solution be denoted by $x^{(i)}(t; r^{(i)}; s^{(i)})$. Then, in the entire interval,

$$x(t) \equiv x^{(i)}(t; r^{(i)}; s^{(i)}), \quad t_i \leq t \leq t_{i+1}, \quad 0 \leq i < N. \tag{5.58}$$

Here, we need to find $r = [r^{(0)}; r^{(1)}; \ldots; r^{(N-1)}]$ and $s = [s^{(0)}; s^{(1)}; \ldots; s^{(N-1)}]$ such that

$$
\begin{aligned}
x^{(i)}(t_{i+1}; r^{(i)}; s^{(i)}) &= r^{(i+1)}, \\
v^{(i)}(t_{i+1}; r^{(i)}; s^{(i)}) &= s^{(i+1)},
\end{aligned}
\qquad 0 \le i < N-1, \qquad (5.59)
$$

and $x^{(0)}(t_0; r^{(0)}; s^{(0)}) = x^0, x^{(N-1)}(t_N; r^{(N-1)}; s^{(N-1)}) = x^e$ (see Fig. 5.5). If we define a vector function F such that

$$
F(r; s) \equiv
\begin{bmatrix}
x^{(0)}(t_1; r^{(0)}; s^{(0)}) - r^{(1)} \\
v^{(0)}(t_1; r^{(0)}; s^{(0)}) - s^{(1)} \\
\vdots \\
x^{(N-2)}(t_{N-1}; r^{(N-2)}; s^{(N-2)}) - r^{(N-1)} \\
v^{(N-2)}(t_{N-1}; r^{(N-2)}; s^{(N-2)}) - s^{(N-1)} \\
x^{(0)}(t_0; r^{(0)}; s^{(0)}) - x^0 \\
x^{(N-1)}(t_N; r^{(N-1)}; s^{(N-1)}) - x^e
\end{bmatrix},
\qquad (5.60)
$$

then we essentially need to determine $(r; s)$ such that $F(r; s) = 0$. The latter can be solved by using a conventional nonlinear equation solver, say, the Newton method.

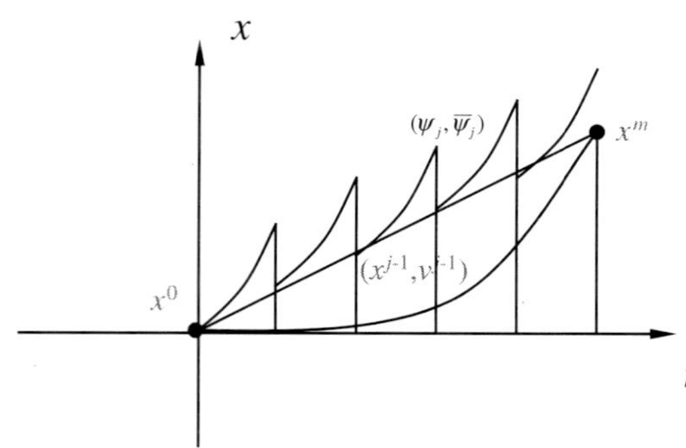

Fig. 5.5. Multiple shooting. The trajectory is divided into a set of subtrajectories, whose initial positions and velocities are adjusted (or determined) so that the resulting trajectories can be pieced together smoothly.

Note that solving the system of equations $F(r; s) = 0$ is not trivial. First, it in fact has N subsystems,

$$x^{(i)}(t_{i+1}; r^{(i)}; s^{(i)}) - s^{(i+1)} = 0,$$
$$v^{(i)}(t_{i+1}; r^{(i)}; s^{(i)}) - r^{(i+1)} = 0, \qquad 0 \le i < N - 1, \quad (5.61)$$

and

$$x^{(0)}(t_0; r^{(0)}; s^{(0)}) - x^0 = 0,$$
$$x^{(N-1)}(t_N; r^{(N-1)}; s^{(N-1)}) - x^{\mathrm{e}} = 0, \qquad (5.62)$$

where each subsystem in turn has $6n$ equations. There are a total of $6nN$ equations. The variables include the initial position and velocity vectors of the solution trajectory in all of the subintervals. Therefore, the total number of variables is also $6nN$.

For proteins, n is typically on the order of 10 000. Therefore, if the interval is divided into 100 subintervals, there will be a system of one million equations with one million variables to solve for. Suppose that each subinterval has a length of 10 ps. A nanosecond trajectory would require the solution of a system of roughly this size. However, the system can be solved more efficiently than it looks.

First, each of the equations in the system has only a small subset of all variables, and therefore the Jacobian of F is very sparse. By exploiting the sparse structure of the problem, the calculations can be significantly reduced. Second, the evaluation of the equations in each subsystem requires the solution of an initial-value problem in the corresponding time subinterval. While evaluating the equations for all of the subsystems would take a substantial amount of computing time, the evaluations of the equations of the subsystems are independent of each other, and can be carried out in parallel on their own subintervals. The latter property makes the multiple-shooting method more scalable and hence more efficient on parallel computers than conventional molecular dynamics simulation schemes.

5.4. Normal Mode Analysis

Another important computational method for the study of protein dynamics is normal mode analysis. When a protein reaches an equilibrium state, the atoms still vibrate around their equilibrium positions, resulting in overall protein structural fluctuations. The fluctuations of the structure can certainly be observed through conventional molecular dynamics simulation, but they can also be studied using a more efficient method such as normal mode analysis.

5.4.1. *Equilibrium state approximation*

The dynamic behavior of a protein of n atoms is described by a system of equations of motion as we have discussed in the previous sections,

$$Mx'' = -\nabla E(x), \tag{5.63}$$

where M is the mass matrix, E is the potential energy function, and x contains the coordinates of the atoms. For the purpose of the following discussions, we put the coordinates of the atoms in a single vector and write $x = \{x_i : i = 1, 2, \ldots, 3n\}$. Assume that the energy at the equilibrium state x^0 is equal to zero. Then, around the equilibrium state, the energy function can be approximated by a second-order Taylor expansion:

$$E(x) \approx \Delta x^{\mathrm{T}} H \Delta x / 2, \quad H = \nabla^2 E(x^0), \tag{5.64}$$

where $\Delta x = x - x^0$. It follows from Eq. (5.63) that

$$M \Delta x'' = -H \Delta x. \tag{5.65}$$

Based on this equation, at equilibrium, a protein structure fluctuates as if the protein is governed by a harmonic potential around the equilibrium state, and the force field becomes approximately linear (see Fig. 5.6). The equation can then be solved analytically through the singular value decomposition of the Hessian matrix H.

5.4.2. *Normal modes*

Let $U \Lambda U^{\mathrm{T}}$ be the singular value decomposition of H, where U is an orthogonal matrix, $UU^{\mathrm{T}} = M$, and $\Lambda = \mathrm{diag}[\Lambda_1, \Lambda_2, \ldots, \Lambda_{3n}]$. Then,

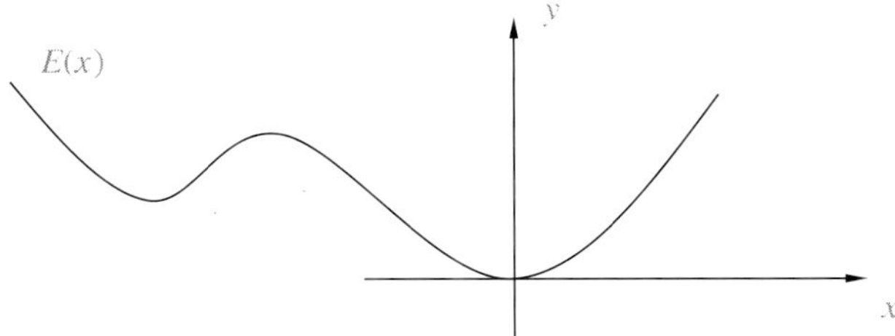

Fig. 5.6. Equilibrium state approximation. At an equilibrium state, i.e. energy minimum, the energy function can be defined by using a quadratic approximation.

Eq. (5.65) becomes,

$$UU^{\mathrm{T}}\Delta x'' = -U\Lambda U^{\mathrm{T}}\Delta x. \tag{5.66}$$

Let $\Delta y = U^{\mathrm{T}}\Delta x$. The equation can further be written as

$$\Delta y'' = -\Lambda \Delta y, \tag{5.67}$$

or equivalently,

$$\Delta y''_j = -\Lambda_j \Delta y_j, \quad j = 1, 2, \ldots, 3n. \tag{5.68}$$

The solutions to the above equations can easily be obtained as

$$\Delta y_j(t) = \alpha_j \cos(\omega_j t + \beta_j), \quad j = 1, 2, \ldots, 3n, \tag{5.69}$$

and then,

$$\Delta x_i(t) = \sum_{j=1}^{3n} U_{i,j}\alpha_j \cos(\omega_j t + \beta_j), \quad i = 1, 2, \ldots, 3n. \tag{5.70}$$

Here, the jth component of the sum is called the jth mode of the motion. The frequency of the jth mode $\omega_j = \Lambda_j^{\frac{1}{2}}$ with α_j and β_j can be determined by the system's initial conditions. Let a set of initial conditions of the system be given as follows:

$$\Delta x_i(0) = 0, \quad \Delta v_i(0) = \Delta v_i^0, \quad i = 1, 2, \ldots, 3n. \tag{5.71}$$

Then,

$$\sum_{j=1}^{3n} U_{i,j}\alpha_j \cos(\beta_j) = 0,$$
$$\sum_{j=1}^{3n} U_{i,j}\alpha_j \omega_j \sin(\beta_j) = \Delta v_i^0, \qquad i = 1, 2, \ldots, 3n. \qquad (5.72)$$

By solving the first set of equations,

$$\sum_{j=1}^{3n} U_{i,j}\alpha_j \cos(\beta_j) = 0, \quad i = 1, 2, \ldots, 3n,$$

for $\alpha_j \cos(\beta_j)$, we obtain

$$\alpha_j \cos(\beta_j) = 0, \quad j = 1, 2, \ldots, 3n. \qquad (5.73)$$

By solving the second set of equations,

$$\sum_{j=1}^{3n} U_{i,j}\alpha_j \omega_j \sin(\beta_j) = \Delta v_i^0, \quad i = 1, 2, \ldots, 3n,$$

for $\alpha_j \sin(\beta_j)$, we obtain

$$\alpha_j \sin(\beta_j) = \omega_j^{-1} \sum_{i=1}^{3n} U_{i,j}\Delta v_i^0, \quad j = 1, 2, \ldots, 3n. \qquad (5.74)$$

By combining Eqs. (5.73) and (5.74), we have

$$\beta_j = \frac{\pi}{2}, \quad \alpha_j = \omega_j^{-1} \sum_{i=1}^{n} U_{i,j}\Delta v_i^0, \quad j = 1, 2, \ldots, 3n. \qquad (5.75)$$

We then see that the amplitude α_j of the motion in the jth mode is inversely proportional to the frequency ω_j of the corresponding mode. Therefore, the larger the frequency ω_j, i.e. the faster the mode, the smaller the amplitude α_j.

5.4.3. *Thermodynamic properties*

Based on the solution to the approximated equation of motion at equilibrium, the potential and kinetic energies of the molecule can be expressed as

$$E_{\mathrm{p}} = \frac{1}{2}\Delta x^{\mathrm{T}} H \Delta x = \frac{1}{2}\Delta y^{\mathrm{T}} \Lambda \Delta y = \frac{1}{2}\sum_{j=1}^{3n} \Lambda_j \Delta y_j^2, \qquad (5.76)$$

$$E_{\mathrm{k}} = \frac{1}{2}\Delta x'^{\mathrm{T}} M \Delta x' = \frac{1}{2}\Delta y'^{\mathrm{T}} \Delta y' = \frac{1}{2}\sum_{j=1}^{3n} \Delta y_j'^2, \qquad (5.77)$$

respectively. Based on the theory of classical dynamics, the time-averaged potential or kinetic energy for each mode should be equal to $k_{\mathrm{B}}T/2$, where k_{B} is the Boltzmann constant and T is the temperature. Therefore,

$$\frac{1}{2}\Lambda_j \langle \Delta y_j^2 \rangle = \frac{1}{2}k_{\mathrm{B}}T, \qquad (5.78)$$

$$\frac{1}{2}\langle \Delta y_j'^2 \rangle = \frac{1}{2}k_{\mathrm{B}}T. \qquad (5.79)$$

It follows from either case that

$$\alpha_j^2 = \frac{2k_{\mathrm{B}}T}{\Lambda_j^{-1}} = \frac{2k_{\mathrm{B}}T}{\omega_j^{-2}}. \qquad (5.80)$$

Two important thermodynamic properties can be calculated from the solution to the approximated equation of motion: the correlations among the dynamic modes and among the atomic fluctuations. The correlation between two given modes can be calculated by using the formula

$$\langle \Delta y_i \Delta y_j \rangle = \lim_{\tau \to \infty} \frac{1}{\tau} \int_0^\tau \Delta y_i(t) \Delta y_j(t)\, \mathrm{d}t, \qquad (5.81)$$

and the correlation between the fluctuations of two atoms by

$$\langle \Delta x_i \Delta x_j \rangle = \lim_{\tau \to \infty} \frac{1}{\tau} \int_0^\tau \Delta x_i(t) \Delta x_j(t)\, \mathrm{d}t. \qquad (5.82)$$

It follows that

$$\langle \Delta y_i \Delta y_j \rangle = \begin{cases} 0, & i \neq j, \\ \alpha_i^2/2, & i = j, \end{cases} \tag{5.83}$$

and

$$\langle \Delta x_i \Delta x_j \rangle = \begin{cases} \frac{1}{2} \sum_{k=1}^{3n} U_{i,k} U_{j,k} \alpha_k^2, & i \neq j, \\ \frac{1}{2} \sum_{k=1}^{3n} U_{i,k} U_{i,k} \alpha_k^2, & i = j. \end{cases} \tag{5.84}$$

In particular,

$$\langle \Delta x_i \Delta x_i \rangle = \frac{1}{2} \sum_{j=1}^{3n} U_{i,j} \alpha_j^2 U_{i,j} = k_{\mathrm{B}} T \sum_{j=1}^{3n} U_{i,j} \Lambda_j^{-1} U_{i,j},$$
$$i = 1, 2, \ldots, 3n, \tag{5.85}$$

and are called the mean square fluctuations of the atoms.

From the above formulas, we see that the atomic fluctuations are collections of motions of different modes, and the modes with smaller singular values make larger contributions to the fluctuations. Therefore, in practice, assuming that the singular values are given in an increasing order, a good approximation to the calculation of the fluctuations can actually be obtained by using only the first few terms in the formulas.

5.4.4. *Gaussian network modeling*

The Gaussian network model is another approach for evaluating the fluctuations of protein structures, especially at the residue level of the protein. The advantage of this approach is that it does not require the availability of an accurate energy function for evaluating the Hessian matrix. The energy function is obtained approximately by forming an elastic network model for the protein with its residue contact matrix. The normal modes are then calculated directly from the singular value decomposition of the contact matrix. This approach is also more efficient than normal mode analysis because it models the system at a coarse level, and fewer modes are required for the estimation.

In a coarse protein model, a residue is represented by a point such as the position of C_α or C_β in the residue, and the protein is considered as a sequence of such points connected with strings. Given such a

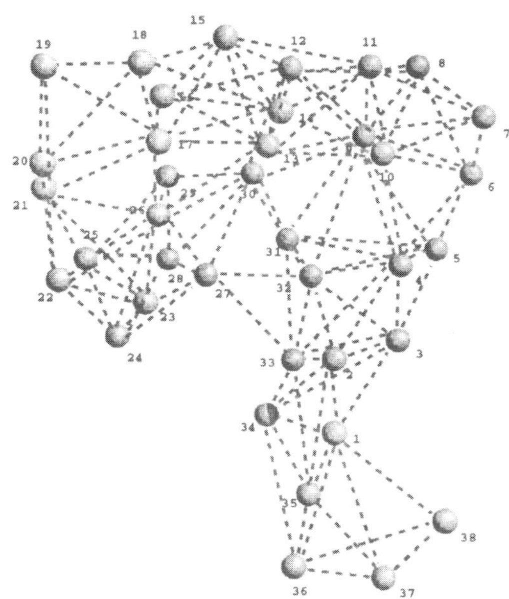

Fig. 5.7. Gaussian network model. The residues in contact are connected with links (like springs).

model, a so-called contact map can be constructed to show how the residues in the protein interact with each other. The map is represented by a matrix with its i,j entry equal to -1 if the residues i and j are within, say, $7\,\text{Å}$ distance, and to 0 otherwise. The contact matrix can be used to compare different proteins. Similar contact patterns often imply structural or functional similarities between proteins (Fig. 5.7).

When a protein reaches its equilibrium state, the residues in contact can be considered as a set of masses connected with springs. A simple energy function can then be defined using the contact matrix of the protein. Suppose that a protein has n residues with n coordinate vectors $x_i = (x_{i1}, x_{i2}, x_{i3}), i = 1, 2, \ldots, n$. Let Γ be the contact matrix for the protein in its equilibrium state:

$$\Gamma_{i,j} = \begin{cases} -1, & \|x_i - x_j\| \le 7\,\text{Å} \\ 0, & \text{otherwise} \end{cases} \quad i \ne j = 1, 2, \ldots, n,$$

$$\Gamma_{i,i} = -\sum_{j=1}^{n} \Gamma_{i,j}, \quad i = j.$$

(5.86)

Then, a potential energy function E for the protein can be defined such that for any vector $\Delta x = (\Delta x_1, \Delta x_2, \ldots, \Delta x_n)^{\mathrm{T}}$ of the displacements of the residues from their equilibrium positions,

$$E(\Delta x) = \frac{1}{2} k \, \Delta x^{\mathrm{T}} \Gamma \Delta x, \tag{5.87}$$

where k is a spring constant. According to statistical physics, the probability for the protein to have a displacement Δx at temperature T should then be subject to the Boltzmann distribution,

$$p_T(\Delta x) = \frac{1}{Z} \exp(-E(\Delta x)/k_{\mathrm{B}}T)$$

$$= \frac{1}{Z} \exp(-k\Delta x^{\mathrm{T}}\Gamma\Delta x/2k_{\mathrm{B}}T), \tag{5.88}$$

where Z is the normalization factor and k_{B} is the Boltzmann constant.

With the above distribution function, we can find the probabilities of the residues fluctuating around their equilibrium positions and, in particular, estimate the mean square fluctuations for the residues, $\langle \Delta x_i^{\mathrm{T}} \Delta x_i \rangle$, $i = 1, 2, \ldots, n$. The estimation requires only some simple linear algebraic calculations. In fact, let the singular value decomposition of Γ be given as $\Gamma = U \Lambda U^{\mathrm{T}}$. Then,

$$E(\Delta x) = \frac{1}{2} k \Delta x^{\mathrm{T}} U \Lambda U^{\mathrm{T}} \Delta x$$

and

$$\langle \Delta x_i^{\mathrm{T}} \Delta x_i \rangle = \frac{1}{Z} \int_{R^{3n}} \Delta x_i^{\mathrm{T}} \Delta x_i \exp\left(\frac{-E(\Delta x)}{k_{\mathrm{B}}T} \right) \mathrm{d}\Delta x$$

$$= \frac{1}{Z} \int_{R^{3n}} \Delta x_i^{\mathrm{T}} \Delta x_i \exp\left(\frac{-k\Delta x^{\mathrm{T}} U\Lambda U^{\mathrm{T}} \Delta x}{2k_{\mathrm{B}}T} \right) \mathrm{d}\Delta x$$

$$= \frac{3k_{\mathrm{B}}T \sum_{j=1}^{n} U_{i,j} \Lambda_j^{-1} U_{i,j}}{k}. \tag{5.89}$$

The above formula is very similar to the formula for calculating the atomic fluctuations in normal mode analysis. In fact, it is not difficult to verify that the residue fluctuations predicted by the

Gaussian network model can also be obtained through normal mode analysis, with an energy function given at the residue level. Let $x^0 = \{x_i^0 : i = 1, 2, \ldots, n\}$ be the equilibrium positions of the residues in the protein. Assume that the energy at x^0 is equal to zero. Then,

$$E(x) = \frac{1}{2}k\,\Delta x^{\mathrm{T}}\Gamma\Delta x \tag{5.90}$$

for $x = x^0 + \Delta x$. With this energy function, the formula for calculating the residue fluctuations as shown in Eq. (5.89) can be derived in the same form from normal mode analysis.

Selected Further Readings

Equation of motion

Arnold VI, Weinstein A, Vogtmann K, *Mathematical Methods of Classical Mechanics*, Springer, 1997.

Goldstein H, Poole CP, Safko JL, *Classical Mechanics*, Addison-Wesley, 2002.

Thornton ST, Marion JB, *Classical Dynamics of Particles and Systems*, Brooks Cole, 2003.

Leach A, *Molecular Modeling: Principles and Applications*, Prentice Hall, 2001.

Haile JM, *Molecular Dynamics Simulation: Elementary Methods*, Wiley, 2001.

Schlick T, *Molecular Modeling and Simulation: An Interdisciplinary Guide*, Springer, 2003.

MaCammon JA, Harvey SA, *Dynamics of Proteins and Nucleic Acids*, Cambridge University Press, 2004.

Brooks III CL, Karplus M, Pettitt BM, *Proteins: A Theoretical Perspective of Dynamics, Structure, and Thermodynamics*, Advances in Chemical Physics, Vol. 71, Wiley, 2004.

Initial-value problem

Brooks BR, Bruccoleri RE, Olafson BD, States DJ, Swaminathan S, Karplus M, CHARMM: A program for macromolecular energy, minimization, and dynamics calculations, *J Comput Chem* 4: 187–217, 1983.

Case D, Darden T, Cheatham T, Simmerling C, Wang J, Duke R, Luo R, Merz K, Wang B, Pearlman B, Crowley M, Brozell S, Tsui V, Gohlke H, Mongan J, Hornak V, Cui G, Beroza P, Schafmeister C, Caldwell J, Ross W, Kollman P, *AMBER 8*, University of California, San Francisco, 2004.

Verlet L, Computer experiments on classical fluids. I. Thermo dynamical properties of Lennard–Jones molecules, *Phys Rev* **159**: 98–103, 1967.

McCammon JA, Gelin BR, Karplus M, Dynamics of folded proteins, *Nature* **267**: 585–590, 1977.

Karplus M, McCammon JA, Protein structural fluctuations during a period of 100 ps, *Nature* **277**: 578, 1979.

McCammon JA, Wolynes PG, Karplus M, Picosecond dynamics of tyrosine side chains in proteins, *Biochemistry* **18**: 927–942, 1979.

Weiner S, Kollman P, Nguyen D, Case D, An all-atom force field for simulations of proteins and nucleic acids, *J Comput Chem* **7**: 230–252, 1986.

Martys N, Mountain R, A velocity Verlet algorithm for dissipative particle dynamics based models of suspensions, *Phys Rev* **E59**: 3733–3736, 1999.

Martyna G, Tuckerman M, Symplectic reversible integrators: Predictor-corrector methods, *J Chem Phys* **102**: 20–22, 1995.

Ryckaert JP, Ciccotti FH, Berendsen HJC, Numerical integration of the Cartesian equations of motion of a system with constraints — Molecular dynamics of n-alkanes, *J Comput Phys* **23**: 327–341, 1977.

Andersen HC, Rattle: A velocity version of the Shake algorithm for molecular dynamics calculations, *J Comput Phys* **52**: 24–34, 1983.

Gunsteren WFV, Karplus M, Effect of constraints on dynamics of macromolecules, *Macromolecules* **15**: 1528–1544, 1982.

Barth E, Kuczera K, Leimkuhler B, Skeel RD, Algorithms for constrained molecular dynamics, *J Comput Chem* **16**: 1192–1209, 1995.

Boundary-value problem

Olender R, Elber R, Calculation of classical trajectories with a very large time step: Formalism and numerical examples, *J Chem Phys* **105**: 9299–9315, 1996.

Elber R, Meller J, Olender R, Stochastic path approach to compute atomically detailed trajectories: Application to the folding of C peptide, *J Phys Chem* **B103**: 899–911, 1999.

Elber R, Ghosh A, Cardenas A, Long time dynamics of complex systems, *Acc Chem Res* **35**: 396–403, 2002.

Elber R, Ghosh A, Cárdenas A, Stern H, Bridging the gap between reaction pathways, long time dynamics and calculation of rates, *Adv Chem Phys* **126**: 93–129, 2003.

Bolhuis P, Chandler D, Dellago C, Geissler P, Transition path sampling: Throwing ropes over mountain passes in the dark, *Annu Rev Phys Chem* **59**: 291–318, 2002.

Chekmarev D, Ishida D, Levy R, Long time conformational transitions of alanine dipeptide in aqueous solution: Continuous and discrete state kinetic models, *J Phys Chem* **B108**: 19487–19495, 2004.

Gladwin B, Huber T, Long time scale molecular dynamics using least action, *ANZIAM J* **E45**: C534–C550, 2004.

Malolepsza E, Boniecki M, Kolinski A, Piela L, Theoretical model of prion propagation: A misfolded protein induces misfolding, *Proc Natl Acad Sci USA* **102**: 7835–7840, 2005.

Ascher U, Mattheij R, Russell R, *Numerical Solution of Boundary Value Problems for Ordinary Differential Equations*, SIAM, 1995.

Deuflhard P, Bornemann F, *Scientific Computing with Ordinary Differential Equations*, Springer, 2002.

Deuflhard P, *Newton Methods for Nonlinear Problems*, Springer, 2004.

Stoer J, Bulirsch R, *Introduction to Numerical Analysis*, Springer, 2002.

Normal mode analysis

Brooks B, Karplus M, Harmonic dynamics of proteins: Normal modes in bovine pancreatic trypsin inhibitor, *Proc Natl Acad Sci USA* **80**: 6571–6575, 1983.

Levitt M, Sander C, Stern PS, Protein normal-mode dynamics: Trypsin inhibitor, crambin, ribonuclease and lysozyme, *J Mol Biol* **181**: 423–447, 1985.

Haliloglu T, Bahar I, Erman B, Gaussian dynamics of folded proteins, *Phys Rev Lett* **B79**: 3090–3093, 1997.

Atilgan A, Durell S, Jernigan R, Demirel M, Keskin O, Bahar I, Anisotropy of fluctuation dynamics of proteins with an elastic network model, *Biophys J* **80**: 505–515, 2001.

Kundu S, Melton JS, Sorensen DC, Phillips Jr GN, Dynamics of proteins in crystals: Comparison of experiment with simple models, *Biophys J* **83**: 723–732, 2002.

Cui Q, Bahar I, *Normal Mode Analysis: Theory and Applications to Biological and Chemical Systems*, Chapman & Hall/CRC, 2005.

Chapter 6

Knowledge-based Protein Modeling

Knowledge-based protein modeling is so named as a contrast to the theory- or experiment-based approaches to protein structure determination. It is called knowledge-based because it attempts to determine protein structures by using mainly the knowledge derived from known protein structures. Two types of structural properties can be derived from known protein structures: those preserved by proteins of similar sequences, structures, or functions; and those shared by all known proteins. The first type can be obtained by comparing the sequences, structures, or even functions of the proteins; while the second can be extracted from the statistical distributions of properties in databases of known protein structures.

6.1. Sequence/Structural Alignment

Proteins with similar sequences may come from the same family of genes and have similar structures and functions. Therefore, by comparing sequences, we may be able to predict the structures or even functions of some unknown proteins. Such an approach to protein modeling is called homology modeling.

It has been estimated that there can be as many as hundreds of thousands of different genes, but they fold into only several tens of thousands of different structures. This implies that there must be many homologous proteins that fold into the same or similar structures. Indeed, the structures of two proteins can be very similar, at least in important regions, even if their sequence identity is as low as 50%. Therefore, homology modeling can usually provide a fairly reasonable model for an unknown protein, if some sequence similarity between the given protein and a protein of known structure can be found. The percentage of identical residues required for a good model is a function of the number of aligned residues. This tends to 25% as the number of aligned residues exceeds 200 (see Fig. 6.1).

In fact, there are proteins with distinct sequences (less than 25% identity) that still fold to the same or similar structures. Such proteins are likely to have similar functions or, in other words, be biologically related, but cannot be identified through sequence comparison. They can be found instead by using so-called structural alignment. With this technique, two proteins are compared in terms of the structural compatibility rather than the sequence compatibility of the amino acid pairs.

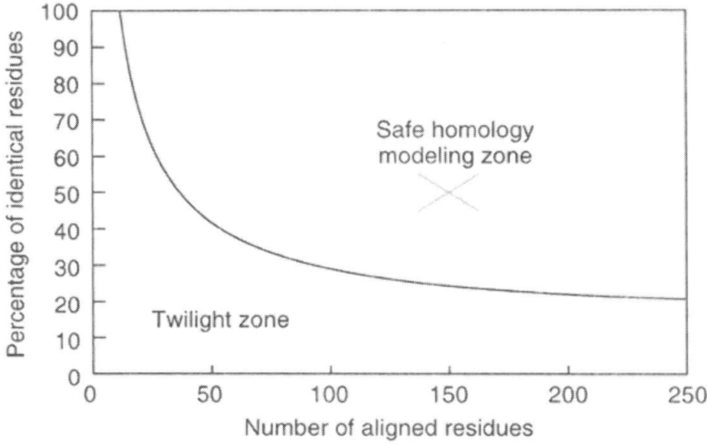

Fig. 6.1. Homology modeling. The requirement on the percentage of identical residues for a good model is a function of the number of aligned residues. It tends to 25% as the number of aligned residues is larger than 200.

6.1.1. *Sequence alignment*

The differences between sequences are caused by genetic mutations such as deletions or substitutions in certain parts of the genes during evolution. Therefore, when comparing two sequences, they need to be properly aligned because they may contain differences due to mutations and thus their similarity may not be revealed. For example, the similarity between ATGG and TCG is not clear unless we assume that, in front of the second sequence, there may be a deletion of one base. With this assumption, the two sequences contain two identical bases, and they can therefore be considered genetically similar.

One of the principles for sequence comparison is that the comparison must be made with the best possible alignment; that is, the similarity between two sequences is defined as the highest similarity that can be identified from all possible alignments of the sequences. As one can imagine, different assumptions can be made as to possible mutations; for example, deletions may occur in any position within the sequences. Enumerating all possible combinations would be difficult if the sequences are long. Fortunately, this can be done efficiently by using a so-called dynamic programming method.

Two proteins can be compared at either the DNA or the amino acid level. At the DNA level, if the original genes encoding the proteins are known, the DNA sequences of the genes are compared. The basic units of the sequences are the four DNA nucleotides: A, T, G, and C. At the amino acid level, the amino acid sequences of the proteins are compared. The basic units of the sequences are the 20 different amino acids, each represented by a letter or word. In either case, a scoring function must be defined to give a score for the matches, mismatches, and deletions between individual units so that an assessment of the overall sequence similarity can be made.

6.1.2. *Shortest path problem*

We first consider a related problem in graph theory. Suppose that we have a directed and ordered graph $G = (V, E, D)$, where V is a set of nodes, $V = \{v_i : i = 1, 2, \ldots, n\}$; E is a set of ordered pairs of nodes corresponding to the directed links among the nodes, $E = \{(v_i, v_j) : v_i, v_j$ in V, and $i < j\}$; and D is a set of distances assigned to the links in E, $D = \{d_{i,j} : (v_i, v_j)$ in $E\}$. The graph is ordered because there are only links from nodes with lower indices to nodes with higher indices. We consider a so-called shortest path problem for this graph. The problem is to find a path from the beginning node v_1 to the ending node v_n so that the sum of the distances along the path, as the length of the path, is the shortest (Fig. 6.2).

If we simply enumerate all possible paths from v_1 to v_n, there will be exponentially many in n, and therefore it is impossible to compare them all if n is large, say $n = 100$. However, we can use a dynamic programming method to obtain the solution much more efficiently. The idea is as follows. First, we find the shortest path from v_1 to v_2, then from v_1 to v_3, and so forth. For v_3, we only need to compare all possible paths from v_1 to v_3. There are only two types of such paths: one from v_1 directly to v_3; and another passing v_2 and then to v_3. The shortest path of the second type is certainly the shortest path from v_1 to v_2 plus the distance from v_2 to v_3. For v_k with any $k > 3$, there must be $k-1$ types of paths from v_1 to v_k: the path from v_1 directly to v_k; and the paths passing v_j and then to v_k, $1 < j < k$. The shortest path from v_1, passing v_j, to v_k must be the shortest path from v_1 to v_j plus the distance from v_j to v_k, and it can therefore be obtained immediately. The shortest path from v_1 to v_k must be the shortest one among all those passing v_j and then to v_k and the one directly going from v_1 to v_k.

For each v_k, we need to make only $k - 1$ comparisons (among all of the shortest paths passing v_j and then to v_k). Therefore, for

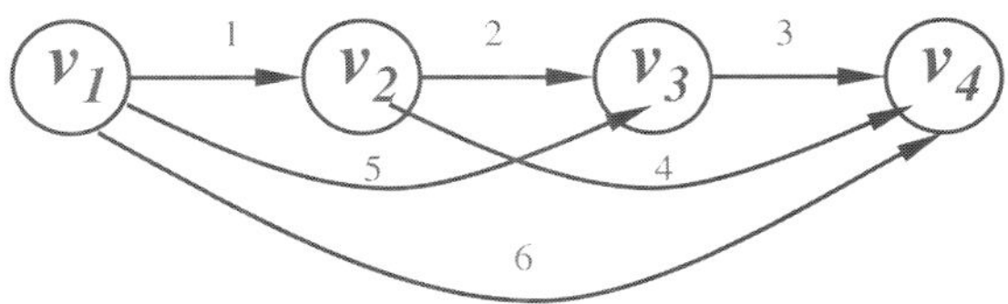

Fig. 6.2. Shortest path problem. Given a set of nodes and links among them, find a path from one node to another with the shortest distance or weight.

$k = 2, 3, \ldots, n$, we need to make $n(n + 1)/2$ comparisons. The total computation time will be in the order of n^2. The algorithm can be written formally as follows.

Solution to the shortest path problem:

```
for j = 1, 2, ..., n
        L_j = d_{1,j}; P_j = 1;
        for i = 2, 3, ..., j - 1
            if L_i + d_{i,j} < L_j
                L_j = L_i + d_{i,j}; P_j = i;
            end
        end
end
P* = v_1 ... v_k v_l ... v_n; k = P_l;
L* = L_n;
```

where, in the end, P^* gives the sequence of the nodes on the shortest path from v_1 to v_n, and L^* is the length of the path.

6.1.3. *Optimal alignment*

The algorithm for solving the shortest path problem for a directed and ordered graph can be extended to the solution of the optimal sequence alignment problem. Suppose we have two sequences of lengths m and n, respectively. We can construct a directed graph with mn nodes arranged in an $m \times n$ matrix. The rows of the matrix correspond to the elements in the first sequence, and the columns to the elements in the second sequence. Each node has three links coming from the left, upper left, and upper nodes. Each also has three links going to the right, lower right, and lower nodes. Each node $v_{i,j}$ corresponds to two elements in the sequences, the ith element in the first sequence and the jth element in the second sequence (Fig. 6.3).

Starting from the first node, for each one, if two corresponding elements are to be compared, the graph makes a transition from this node to the node at its lower right with an assigned score of 0 or 2 on the corresponding link. Here, if the two elements are identical, the

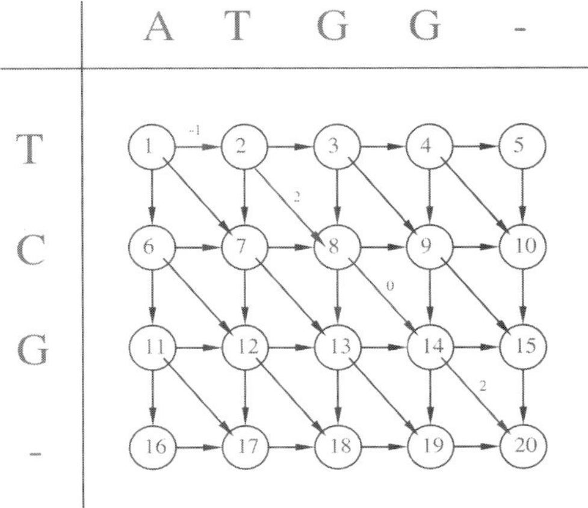

Fig. 6.3. DNA sequence alignment. An alignment between two sequences corresponds to a path from the first and last nodes in the constructed graph.

score is 2; otherwise, it is 0. If an element in the first sequence is to be compared with one in the second sequence next to the current element, the graph makes a transition from the current node to the node directly to its right with an assigned score of -1 on the corresponding link. Similarly, if an element in the second sequence is to be compared with one next to the current element in the first sequence, the graph makes a transition from the current node to the node below with an assigned score of -1 on the corresponding link.

With the above construction, we have a directed and ordered graph. With a sequence of transitions from the first node to the last node, we can make an alignment for the elements in the two sequences. If we add the scores of the transitions for the alignment, we obtain a measure of the alignment, and the highest score gives the best alignment between the two sequences. From a graph-theoretic point of view, each alignment corresponds to a path from the first node to the last node. The score for the alignment is the length of the path, and the best alignment corresponds to the longest path from the first to the last node.

Based on the discussion in the previous section, the longest path problem for a directed and ordered graph can be solved in the same fashion as solving the shortest path problem. The computing time is in the order of the square of the number of nodes in the graph. For the graph constructed for sequence alignment, the total number of nodes is *mn*. Therefore, by using an algorithm similar to the one described in the previous section, the problem of finding the best alignment for two given sequences of lengths m and n can be solved in an order of $(mn)^2$ computing time.

However, for sequence alignment, it is not necessary to repeat the second loop of the algorithm for all previous nodes because there are only three previous nodes connected to the current node: the left, upper left, and upper nodes. In fact, it needs to be repeated only three times for the three connected nodes. As a result, the algorithm can be reduced to require only an order of *mn* computing time, as described below.

Solution to the sequence alignment problem:

$$\text{for } j = 1, 2, \ldots, mn$$
$$L_j = d_{1,j}; \ P_j = 1;$$
$$\text{for } i = \text{left}(j), \text{ upper left}(j), \text{ upper}(j)$$
$$\text{if } L_i + d_{i,j} > L_j$$
$$L_j = L_i + d_{i,j}; \ P_j = i;$$
$$\text{end}$$
$$\text{end}$$
$$\text{end}$$
$$P^* = v_1 \cdots v_k v_l \cdots v_{mn}; \ k = P_l;$$
$$L^* = L_{mn};$$

where, in the end, P^* gives the sequence of nodes on the longest path from v_1 to v_{mn}, and L^* is the length of the path. Let $P = v_1 \cdots v_k \cdots v_{mn}$. Then, given P, we can immediately obtain an alignment between the two given sequences, with the first one in the order of

$$s_{i_1} \cdots s_{i_k} \cdots s_{i_{mn}}$$

and the second one in the order of

$$s_{j_1} \cdots s_{j_k} \cdots s_{j_{mn}},$$

where $k = (i_k - 1)n + j_k$. Note that if $i_k = i_{k+1}$, then s_{i_k} is in fact a space, meaning that the first sequence has a deletion at position k. Similarly, if $j_k = j_{k+1}$, then s_{j_k} is a space, meaning that the second sequence has a deletion at position k (Figs. 6.4 and 6.5).

	A	C	D	E	F	G	H	I	K	L	M	N	P	Q	R	S	T	V	W	Y
A	5	-2	0	1	-2	0	0	-1	0	-1	0	0	1	0	-1	1	0	0	-2	-2
C	-2	8	-2	-3	-3	-2	0	-2	-3	-3	0	-2	-3	-3	-2	-1	-1	-2	-1	-2
D	0	-2	5	2	-2	0	1	-3	0	-2	-1	2	0	1	-2	0	0	-2	-3	-2
E	1	-3	2	5	-3	0	-1	-2	1	-2	-2	1	1	2	0	1	1	-1	-2	-1
F	-2	-3	-2	-3	6	-3	1	0	-3	2	2	-3	-2	-3	-2	-1	-2	0	3	3
G	0	-2	0	0	-3	5	-1	-2	0	-2	-2	0	0	-1	0	0	-1	-1	-2	-3
H	0	0	1	-1	1	-1	5	-1	1	-1	0	1	0	1	2	0	1	-1	0	1
I	-1	-2	-3	-2	0	-2	-1	5	-2	2	2	-2	-2	-3	-2	-1	0	2	0	0
K	0	-3	0	1	-3	0	1	-2	5	-1	-2	1	0	1	2	0	0	-1	-2	-2
L	-1	-3	-2	-2	2	-2	-1	2	-1	5	3	-2	-2	0	-1	-1	0	2	0	0
M	0	0	-1	-2	2	-2	0	2	-2	3	5	-1	-2	0	-2	-1	0	1	-2	-1
N	0	-2	2	1	-3	0	1	-2	1	-2	-1	5	-2	1	0	2	0	-2	-3	-1
P	1	-3	0	1	-2	0	0	-2	0	-2	-2	-2	8	0	0	0	0	-1	-3	-3
Q	0	-3	1	2	-3	-1	1	-3	1	0	0	1	0	5	2	1	0	-1	-1	-2
R	-1	-2	-2	0	-2	0	2	-2	2	-1	-2	0	0	2	5	-1	0	-1	0	-1
S	1	-1	0	1	-1	0	0	-1	0	-1	-1	2	0	1	1	5	2	-1	0	0
T	0	-1	0	1	-2	-1	1	0	0	0	0	0	0	0	0	2	5	0	-1	-2
V	0	-2	-2	-1	0	-1	-1	2	-1	2	1	-2	-1	-1	-1	-1	0	5	-1	0
W	-2	-1	-3	-2	3	-2	0	0	-2	0	-2	-3	-3	-1	0	0	-1	-1	6	3
Y	-2	-2	-2	-1	3	-3	1	0	-2	0	-1	-1	-3	-2	-1	0	-2	0	3	6

Fig. 6.4. Residue exchange matrix. A score is assigned for every pair of residues when they are aligned.

	V	A	T	T	P	D	K	S	W	L	T	V
A	0	5	0	0	1	0	0	1	-2	-1	0	0
S	-1	1	2	2	0	0	0	5	0	-1	2	-1
T	0	0	5	5	0	0	0	2	-1	0	5	0
P	-1	1	0	0	8	0	0	0	-3	-2	0	-1
E	-2	1	1	1	1	2	1	1	-2	-2	1	-1
R	-1	-1	0	0	0	-2	2	1	0	-1	0	-1
A	0	5	0	0	1	0	0	1	-2	-1	0	0
S	-1	1	2	2	0	0	0	5	0	-1	2	-1
W	-1	-2	-1	-1	-3	-3	-2	0	6	0	-1	-1
L	2	-1	0	0	-2	-2	-1	-1	0	5	0	2
G	-1	0	-1	-1	0	0	0	0	-2	-2	-1	-1
T	0	0	5	5	0	0	0	2	-1	0	5	0
A	0	5	0	0	1	0	0	1	-2	-1	0	0

Fig. 6.5. Protein sequence alignment. All possible alignments for two given sequences can be traced in a matrix, as shown in the figure. A score is given for every pair of aligned residues using the residue exchange matrix. The alignment with the highest total score is selected as the optimal alignment.

6.1.4. *Structural alignment*

There exist proteins with similar structures or functions, but not similar sequences. The similarity between such proteins thus cannot be identified through sequence comparison. Structural alignment is a more general protein comparison method that compares proteins in terms of their structural compatibilities, i.e. how similar the corresponding pairs of residues in the proteins are in terms of their structural properties such as the neighborhood structures of the residues.

In order to make a structural comparison, a protein of known structure is typically represented as a sequence, corresponding to its residue sequence, of structural properties for the residues, including for example the buried area of the side chain, the exposure of the side chain to the solvent, and the local secondary structure. The residues of the protein of unknown structure are then aligned and assigned to the structural properties of the known protein, and assessed on their structural compatibilities. The alignment that yields the best structural compatibility is selected.

Structural compatibility can be defined differently if a different set of structural properties is considered. Once a set of structural properties is selected, the compatibility of each residue with the properties can be determined based on the distributions of the properties for the residue in known proteins. A high compatibility score can be assigned if the residue has the properties in many of its occurrences in the known proteins.

For example, one of the approaches to evaluating structural compatibility — proposed originally by Bowie *et al.* (1991) — is to define a structural environment for each residue with three structural properties including the buried area of the side chain, the exposed area of the side chain to solvent, and the local secondary structure of the residue. Each of these properties is divided further into several different categories. Together, they make 18 different combinations of structural properties as the possible structural environments for the residue. For each of the 20 residues, the probability for the residue to have a particular type of structural environment can be estimated by searching through a database of known protein structures and collecting all of the cases where the residue is found in the specified structural environment.

Let the residue type be i, and the environment type be j. The structural compatibility of residue i in environment j is defined by the formula

$$s(i, j) = \log \frac{p(i, j)}{p(i)},$$

where $p(i, j)$ is the probability of residue i in environment j, and $p(i)$ the probability of residue i in all possible environments. Once the scoring function is defined, the sequence of residues of a given protein can be aligned and compared with the sequence of structural environments of a known protein, by using the same dynamic programming algorithm for sequence alignment as given in the previous section.

6.2. Fold Recognition/Inverse Folding

Knowledge-based protein modeling includes two major subjects: fold recognition and inverse folding. Fold recognition refers to finding a known protein fold (from a database of known protein structures) as a model for a given protein sequence, while inverse folding is the opposite — finding a protein sequence (from a database of known protein sequences) that may form a structure similar to a given protein fold.

6.2.1. *Fold recognition*

Fold recognition typically consists of three steps. First, search through a database of known protein folds and find a set of candidate structures for the given sequence by using sequence or structural alignment. Second, thread the structural components of the candidate proteins on the given sequence and form a set of initial models. Third, assess the threaded structures and choose the best one.

The first and most essential step requires executing a sequence or structural alignment algorithm. The second step involves the assembly of structural components such as secondary structure formation, rigid body movement, loop modeling, etc. In the third step, a scoring function or, sometimes, an energy function, is required to evaluate the structures. The one with the highest score or the lowest energy may be selected as the final model for the protein.

The structure obtained from fold recognition may be further refined by using an optimization method with the scoring function or the energy function as the objective function. For example, by using a classical mechanical potential energy function, the structure can be refined to a more energetically favorable state. In any case, without further experimental evidence, the structures predicted via fold recognition are still theoretical models and have to be evaluated, analyzed, and utilized with caution.

6.2.2. *Inverse folding*

Inverse folding plays an essential role in protein design. For example, if we want to develop a novel protein with a desired structure and function, we need to find the sequences that may indeed fold into the targeted structure. On the other hand, inverse folding can also help to find possible sequences that have the same or similar structures, and thereby make genetic classifications for the sequences according to their structural or even functional similarities.

Inverse folding has three steps. First, search through a database of known protein sequences and find a set of candidate sequences for the given structure by using sequence or structural alignment. Second, thread the structural components on the given structure and form a set of initial models for the found sequences. Third, make an assessment of the threaded structures and choose the best one (Fig. 6.6).

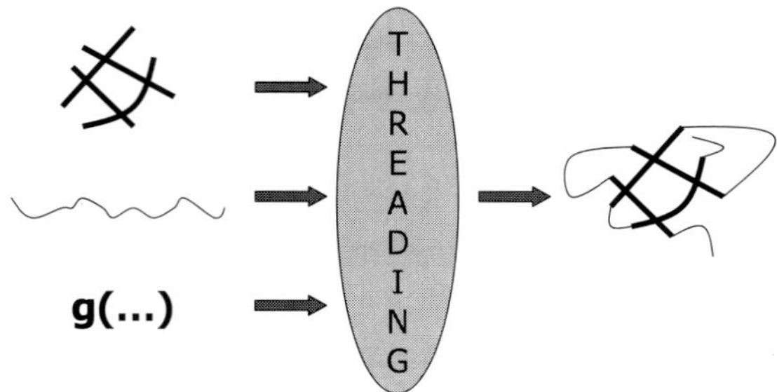

Fig. 6.6. Inverse folding. Given a structure and a sequence, find the best sequence–structure match between them by making the best possible alignments for the core structural fragments.

The first and, again, most essential step requires executing a sequence or structural alignment algorithm. The second step involves the assembly of structural components such as secondary structure formation, rigid body movement, loop modeling, etc. In the third step, the structures are evaluated and compared against the given structure. The sequences with the folds most closely correlated with the given structure are identified.

Inverse folding and fold recognition share some common computational components such as alignment and threading. Alignment has been discussed in the previous section. By using different alignment methods, different approaches to fold recognition and inverse folding may be implemented. Threading involves detailed work on choosing appropriate structural components for different parts of sequences. The difficult part of the work is the arbitrary or combinatorial nature of the choices to be made.

6.2.3. *Scoring functions*

The scoring function is essential to the evaluation of a modeled structure, either in fold recognition or in inverse folding. A scoring function can be an energy function such as those used in protein force field calculations. Based on the general thermodynamic assumption of protein folding that a native protein fold adopts a conformation with the lowest possible energy, a low-energy structure is always preferred among model candidates in fold recognition or inverse folding.

The scoring function can also be defined statistically in terms of some average properties of known protein structures. The score is then, in a certain sense, the probability of the structure to be the native structure as compared with known protein structures. Such a function is called the statistical scoring function or statistical potential. The definition of this function usually requires a complete search of the database of known protein structures as well as statistical estimates on various structural properties.

An example of statistical potential is the distance-based mean force potential by Sippl and Weitckus (1992): for a given pair of atoms or residues of particular types, the distances between the two atoms or residues can be collected from all known protein structures. The collected distances are subject to a certain statistical distribution. The latter can be used to define a statistical potential. For a given protein, the potentials for all of the pairs of atoms or residues can be summed up. The total potential can then be defined as the energy for the entire protein.

Another example of statistical potential is the so-called contact potential by Miyazawa and Jernigan (1985): a pair of residues is said to be in contact if their distance is within a certain range, say $7\,\text{Å}$. The probability for a particular pair of residues to be in contact can be estimated by sampling all such pairs of residues in the database of known protein structures. Then, a contact potential can be defined by using the log of the corresponding contact probability, and the contact energy for the entire protein can be defined by the sum of the contact potentials for all pairs of residues.

6.2.4. *Complexities of threading*

The whole process of aligning a sequence to a structure and thereby guiding the spatial placement of the amino acid sequence is referred to as threading. Various threading algorithms have been developed, some running in polynomial time, while others require exponentially many steps in the worst case scenario. If the alignment is based not only on the individual amino acid types but also on their structural environments, and if gaps of arbitrary lengths are also allowed between sequence segments, an optimal threading has proven to be NP-hard.

In a threading problem, a graph $G = (V, E)$ can be used to represent the interactions among the amino acids in the sequence, where $V = \{v_k : k = 1, 2, \ldots, n\}$ is a set of vertices corresponding to the amino acids in the sequence, and $E = \{(v_i, v_j), \text{ for some } v_i, v_j \text{ in } V\}$ is a set of links representing the relationships among the amino acids. A set of integers $t = \{t_k : k = 1, 2, \ldots, m\}$ can also be used to represent the alignment between the core elements in the template structure and the sequence segments in the sequence, meaning that the core element C_i in the template structure is assigned to the sequence segment starting at position t_i, for $i = 1, 2, \ldots, m$, assuming that there are m core elements

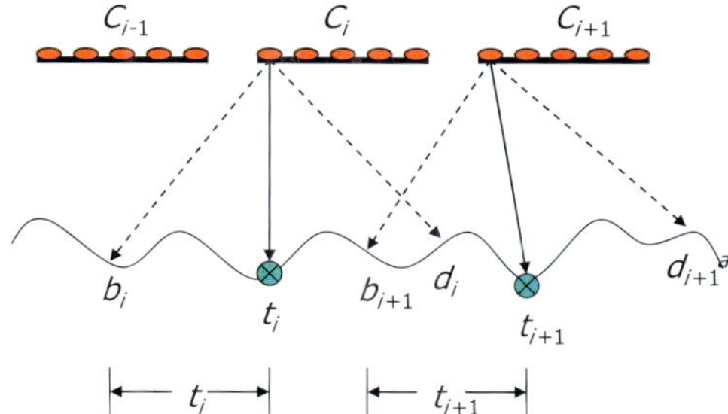

Fig. 6.7. Threading. The core element C_i in the template structure is assigned to the sequence segment started at position t_i. The assigned position must satisfy certain restraints. For example, t_{i+1} must be greater than or equal to $t_i + c_i$, where c_i is the length of C_i.

in the template structure. The assigned position must of course satisfy certain structural restraints. For example, t_{i+1} must be greater than or equal to $t_i + c_i$, where c_i is the length of C_i (Fig. 6.7).

For each threading, a scoring function σ can be defined to calculate the scores for the vertices, links, and gap segments. The scores on the vertices show the matches between the amino acids and the spatial positions. The scores on the links show the compatibilities of the alignments of related amino acids. The scores on the gap segments represent some gap penalty on the entire threading.

Lathrop and Smith (1996) showed that a three-satisfiability problem, which is a classical NP-complete problem, can always be reduced to a protein threading problem represented by a threading graph G, an alignment t, and a scoring function σ. Therefore, the protein threading problem must be NP-hard; otherwise, the three-satisfiability problem could be solved in polynomial time through the solution of the protein threading problem.

6.3. Knowledge-based Structural Refinement

Protein structures determined by either experimental or theoretical techniques are not always as accurate as desired. Further refinement of the structures, including human intervention, is required and sometimes critical. Therefore, the development of an efficient refinement technique is important; and as more and more structures are determined, the need becomes even more urgent, as the CASP (Critical Assessment of protein Structure Prediction) Prediction Center explained in the call for a structure refinement competition in spring 2006.

6.3.1. *Deriving structural constraints*

Here, we examine a knowledge-based approach for protein structure refinement and, in particular, the approach of deriving distance constraints from databases of known protein structures for structure refinement. In this approach, the distributions of the distances of various types in known protein structures are calculated and used to obtain the most probable ranges or the mean force potentials for the distances. The range restrictions or the mean force potentials on the distances are then imposed or applied to the structures to be refined so that more plausible models can be built.

It is intuitive to extract information on interatomic distances based on their distributions in known protein structures. Miyazawa and Jernigan (1985) and Sippl and Weitckus (1992) first studied such distributions, and used them to build statistical potentials for native contact analysis and fold recognition; the idea was later adopted for X-ray structure refinement. Work on the development of mean force potentials for dihedral angles had a similar nature. This idea, however, has not been fully developed into a general approach for structure refinement, though with the increasing number of determined structures, such an approach actually becomes more and more feasible.

6.3.2. *Distance distributions*

A distance can be specified by the types of atoms the distance corresponds to, the types of residues the atoms are associated with, and the number and types of residues separating the two associated residues. We assume that the two atoms of the distance are not in the same residue. Such distances may be called cross-residue, interatomic distances (Fig. 6.8). Let A_1 and A_2 be the types of two corresponding atoms separated by the distance, R_1 and R_2 the types of the two

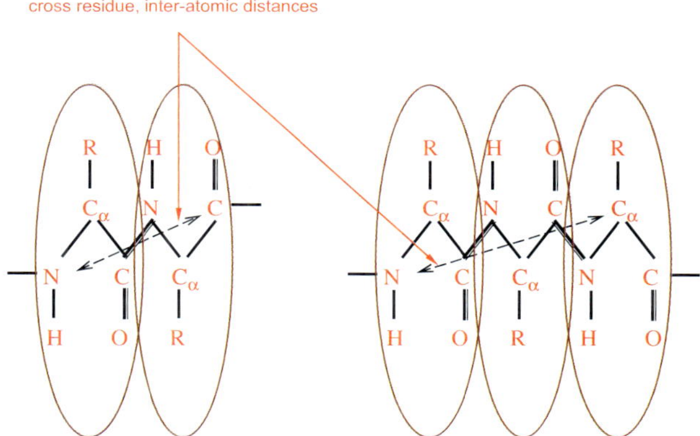

Fig. 6.8. Cross-residue distances. The distances are specified by the types of the two atoms they connect to, the types of residues the two atoms are associated with, and the types of residues separating the two end residues in sequence.

associated residues, and $S_1, S_2, \ldots, S_N$ the types of a maximum N separating residues. Then, the distribution of the distance can be represented by a function $P[A_1, A_2, R_1, R_2, S_1, S_2, \ldots, S_N]$, which can be constructed as follows.

First, all of the distances of this particular type in a database of known protein structures are collected and grouped into a set of uniformly divided distance intervals $[D_i, D_{i+1}]$, where $D_i = 0.1 \times i$ Å, $i = 0, 1, \ldots, n - 1$. The function value $P[A_1, A_2, R_1, R_2, S_1, S_2, \ldots, S_N](D)$ for any distance D in $[D_i, D_{i+1}]$ can then be defined as the number of distances in $[D_i, D_{i+1}]$ divided by the total number of distances in all $[D_i, D_{i+1}], i = 0, 1, \ldots, n - 1$ (Fig. 6.9).

Based on the database distribution of a given type of distance, a possible range of distances can be extracted by using the mean minus and plus a few standard deviations of the distance as the lower and upper bounds, respectively. In general, the lower and upper bounds l and u of distance D of type $[A_1, A_2, R_1, R_2, S_1, S_2, \ldots, S_N]$ can be defined as

$$l[A_1, A_2, R_1, R_2, S_1, S_2, \ldots, S_N] = \mu - k\sigma$$

and

$$u[A_1, A_2, R_1, R_2, S_1, S_2, \ldots, S_N] = \mu + k\sigma,$$

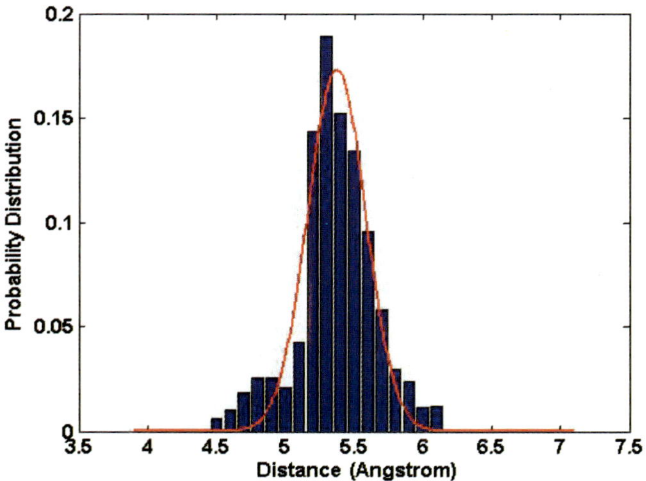

Fig. 6.9. Distribution of distances. A range constraint for the distances may be derived by restricting the distances in the most populated range, say in between mean minus and plus two standard deviations. Or, a mean force potential may be defined for the distances based on the distribution of the distances.

respectively, where μ and σ are the mean and the standard deviation, and k is a constant. For example, to make the range sufficiently large to cover the most probable values of the distance, two standard deviations may be used, in which case k should be equal to 2.

6.3.3. *Mean force potentials*

While easy to apply, the extracted distance bounds not only exclude the possibilities of some distances outside the bounds, but also treat all of the distances within the bounds equally. In fact, the distances are distributed nonuniformly both inside and outside the bounds, and some certainly have higher probabilities than others. Such information can be valuable for the proper choice of distances for a structure. In order to incorporate all of the probability information into the structure refinement, alternatively, we can use the distribution functions of the distances to construct a set of distance-based mean force potentials. Let E be the mean force potential for distance D of type

$[A_1, A_2, R_1, R_2, S_1, S_2, \ldots, S_N]$. Then, E can be defined such that

$$E[A_1, A_2, R_1, R_2, S_1, S_2, \ldots, S_N](D)$$
$$= kT \log P[A_1, A_2, R_1, R_2, S_1, S_2, \ldots, S_N](D),$$

given the distribution function of the distance $P[A_1, A_2, R_1, R_2, S_1, S_2, \ldots, S_N](D)$.

6.3.4. *Structure refinement*

The database-derived distance constraints and mean force potentials have been tested by Cui *et al.* (2005) and Wu *et al.* (2007) for refining nuclear magnetic resonance (NMR)-determined structures and structures obtained from comparative modeling. The results showed that the structures were improved significantly after refinement. The improvements were measured with commonly used modeling standards including ensemble root mean square deviation (RMSD) values, potential energies, Ramachandran plots, etc. In any case, the structures refined with the database-derived constraints (or potentials) are knowledge-based theoretical models that should agree with the experimental results. Therefore, the experimental constraints should always be respectfully considered so that the refined structures are improvements, not changes, of the experimentally determined structures. While the improved parts may result from applying the best knowledge on interatomic distances to structures, they may still require further justification.

6.4. Structural Computing and Beyond

Computing plays an essential role in protein modeling, from data collection to visualization, from structural analysis to classification, and from dynamic simulation to model building and prediction. The development of modern computing technology makes it possible to carry out these tasks; protein modeling could otherwise never have advanced to today's levels and scales.

In addition, with the increasing number of structures being determined, the study of protein structure and its connection to the biological function of the protein has shifted from research on individual proteins to families of proteins or even genome-wide studies. The latter, more reliant on structural computing and information processing, promises even greater theoretical and technological advancements in protein modeling.

6.4.1. *Structural bioinformatics*

Structural bioinformatics, as an emerging field of protein modeling, focuses on the representation, storage, retrieval, analysis, and display of the structural information of proteins. In contrast to experimental or even conventional computational approaches to protein modeling, structural bioinformatics attempts to develop general purpose methods for processing structural information and apply these methods in order to solve structure-related problems.

The development of the field of structural bioinformatics arises from the urgent need to process and analyze the vast amounts of structural data generated from experimental and theoretical studies. The data includes not only the raw X-ray diffraction and NMR spectral data, but also the structural data of known proteins. As more and more proteins are determined, the latter have provided a wealth of structural information that can be used for more complete structural analysis, classification, and modeling.

Structural information is more complicated and difficult to process than sequence information, because it is two- or three-dimensional data and the relationships among different pieces of information are nonlinear. Structures are related more through physics as a result of physical interactions among their constituent components, for example, the atoms or the residues. On the other hand, sequences are more of biological relations, although sequences are connected by chemical bonds.

The conventional X-ray crystallography and NMR data processing and model building are certainly structural bioinformatics activities.

However, what marks structural bioinformatics as an informatics field is the need for representing, storing, and processing the structural data of proteins. A protein can be represented as a linear chain of amino acids, a collection of secondary structures, or, more accurately, a tertiary structure defined at the atom or residue level. The structural data are typically stored in a database, such as the Protein Data Bank (PDB), with each structure described in a structural file. The structural data can be processed for studies with many different purposes.

Structural data can be analyzed or classified according to the structural or functional properties they share by using, for instance, structural alignment. They can also be studied statistically, through database searching, to obtain the distributions of certain important properties in known proteins. The graphical display of structural data is also essential for visualizing and therefore better understanding the various structural features of proteins. A great use of structural data is to provide structural templates for proteins of unknown structures, as we have discussed in the previous sections.

Research on structural bioinformatics is not restricted to processing structural data. It also includes activities such as structure determination, normal mode analysis, and dynamics simulation, as these activities also involve data collection, analysis, and processing, while at the same time producing additional data, such as the protein dynamic trajectories, that are just as important as structural data for understanding the structures and functions of proteins.

6.4.2. *High-performance computing*

Protein modeling is one of the most intensive fields computationally. There are problems for which solutions cannot be obtained without high-performance computing. For example, even with a parallel high-performance computer, it still takes several weeks or months to fold a small protein. Therefore, supercomputer vendors have worked for years to produce machines that are targeted for protein structure and dynamics applications.

A conventional computer has a single processor. In general, the instructions must be executed or processed one at a time, even if they are independent of each other and can be carried out in parallel. In fact, the computation cannot be sped up very much even if the central processing unit (CPU) speed is increased, because many instructions with memory access require many computing cycles and will slow down or block the entire computation.

In order to remove the computational bottleneck of serial execution of instructions on conventional architectures, parallel high-performance computers have been invented, including vector machines developed in the early 1960s, shared-memory machines in the 1970s, and distributed-memory machines in the 1980s. These architectures are capable of parallel execution of instructions in one way or another, depending on their CPU design and memory allocation. They are certainly harder to program, but much more powerful than conventional serial architectures.

Vector machines have CPUs capable of processing a vector of instructions in one or a few CPU cycles. Such architectures are efficient for calculations such as vector or matrix operations, which routinely require the processing of a large set of independent data. However, the data sets must be arranged properly in the CPU vectors so that they can be processed in parallel. Modern vector machines have well-developed operating systems and compilers so that users can program the machines in almost the same way as on conventional serial machines. The vector operations are arranged completely automatically by the systems.

Shared-memory architectures typically have several tens and sometimes hundreds of processors capable of independently executing instructions in parallel. Ideally, if there are 100 processors, the achievable speedup for executing a program can be as high as 100 times. However, a computer program not only executes calculation instructions, but also fetches or stores data in the memory. The shared-memory architectures are designed so that all of the processors share one large memory unit. In this way, they can all visit the same memory at any time they need, and they can be programmed in almost the same way as on conventional serial architectures, except for some necessary synchronization among the processors.

The shared-memory architectures cannot have too many processors because otherwise the processors will have a hard time accessing the memory due to memory traffic. To overcome this difficulty, the distributed-memory architectures were developed so that each processor has its own memory unit and does not have to compete for memory with other processors. In this way, there is in principle no limitation on the number of processors that can be included in the system. However, programming models for distributed-memory machines are more complicated than those for shared-memory machines, and data not local to a processor have to be transferred from elsewhere before the processor can process them. This always introduces additional overheads for data movement or, in other words, processor communication.

6.4.3. *Structural genomics*

The study of structural genomics is one of the major motivations for the development of the field of structural bioinformatics. The purpose of structural genomics is to systematically study the structures of all the proteins encoded in a whole genome. Since the completion of genome sequencing for humans and many other species, structural genomics has become an important and feasible postgenome research subject that is pursued as actively as ever.

Structural genomics aims to study the genes in a given genome and, in particular, the structures of the proteins they encoded. While genome sequencing attempts to identify all of the genetic sequences in a genome and obtain a complete picture of the sequence space, the purpose of structural genomics is to find the structures of the proteins in a genome and obtain the mapping between the sequence space and the structural space of the genome. The latter will be able to provide a complete description of the correlations between sequences and structures, as well as the relationships within the sequences and structures themselves.

The study of structural genomics is a joint effort between experimental and theoretical structural biologists. Given the limitations of experimental approaches for the determination of all required structures, the theoretical approaches — in particular, the homology modeling approach — have been employed to obtain more structures to fill the missing elements in the structural space. The theoretical approaches also rely on the structures determined by the experimental approaches, because they usually predict structures based on the sequence or structural similarities between known and unknown structures.

6.4.4. *Biocomplexes and biosystems*

Proteins are often combined with other macromolecules to form larger protein complexes, and also interact with each other to maintain certain functions or organizations that are required by the biological systems with which they are associated. Therefore, protein complexes and systems biology are subjects of great importance to the study of proteins and their structures. Conversely, the revelation of protein functions and structures is essential for the understanding of higher-level mechanisms of protein complexes and biological systems.

In recent years, systems biology has become an active research subject in computational biology. This is partially due to the rapid development of molecular biology, biotechnology, and genetics so that the understanding of biological systems at a molecular biology level becomes possible. It is also because of the great mathematical and computational challenges present in the modeling and simulation of biological systems. Along this line, how structural information can be utilized for the study of biology systems is certainly an important question to be further explored through research on protein structures.

A biological system can be analyzed by building a network of biological interactions and studying the key metabolic processes, by comparing with other systems and recognizing conserved system functions, and by building accurate mathematical models and identifying important biological pathways. The study of biological functions at a systems level and the understanding of various biological systems including cells, tissues, organs, etc. are certainly the ultimate goals of many biological and medical research activities including computational structural biology.

Selected Further Readings

Sequence/structural alignment

Jones NC, Pevzner PA, *An Introduction to Bioinformatics Algorithms*, The MIT Press, 2004.

Mount DM, *Bioinformatics: Sequence and Genome Analysis*, Cold Spring Harbor Laboratory Press, 2004.

Xiong J, *Essential Bioinformatics*, Cambridge University Press, 2006.

Taylor WR, Orengo CA, Protein structure alignment, *J Mol Biol* **208**: 1–22, 1989.

Orengo CA, Taylor WR, SSAP: Sequence structure alignment program for protein structure comparison, *Methods Enzymol* **266**: 617–635, 1996.

Holm L, Sander C, Mapping the protein universe, *Science* **273**: 595–602, 1996.

Fold recognition/threading

Bowie JU, Lüthy R, Eisenberg D, A method to identify protein sequences that fold into a known three-dimensional structure, *Science* **253**: 164–169, 1991.

Bryant SH, Lawrence CE, An empirical energy function for threading protein sequence through the folding motif, *Proteins* **16**: 92–112, 1993.

Šali A, Blundell TL, Comparative protein modeling by satisfaction of spatial restraints, *J Mol Biol* **234**: 779–815, 1993.

Fischer D, Rice D, Bowie JU, Eisenberg D, Assigning amino acid sequences to 3-dimensional protein folds, *FASEB J* **10**: 126–136, 1995.

Krogh A, Brown M, Mian IS, Sjölander K, Haussler D, Hidden Markov models in computational biology, *J Mol Biol* **235**: 1501–1531, 1994.

Lathrop RH, Smith TF, Global optimum protein threading with gapped alignment and empirical pair score functions, *J Mol Biol* **255**: 641–665, 1996.

Russell RB, Saqi MAA, Sayle RA, Bates PA, Sternberg MJE, Recognition of analogous and homologous protein folds: Analysis of sequence and structure conservation, *J Mol Biol* **269**: 423–439, 1997.

Miyazawa S, Jernigan RL, Estimation of effective inter-residue contact energies from protein crystal structures: Quasi-chemical approximation, *Macromolecules* **18**: 534–552, 1985.

Miyazawa S, Jernigan RL, Residue–residue potentials with a favorable contact pair term and an unfavorable high packing density term for simulation and threading, *J Mol Biol* **256**: 623–644, 1996.

Sippl MJ, Calculation of conformational ensembles from potentials of mean force, *J Mol Biol* **213**: 859–883, 1990.

Sippl MJ, Weitckus S, Detection of native-like models for amino acid sequence of unknown three-dimensional structure in a database of known protein conformations, *Proteins* **13**: 258–271, 1992.

Knowledge-based structural refinement

Kuszewski J, Gronenborn AM, Clore GM, Improving the quality of NMR and crystallographic protein structures by means of a conformational database potential derived from structure databases, *Protein Sci* **5**: 1067–1080, 1996.

Rojnuckarin A, Subramaniam S, Knowledge-based potentials for protein structure, *Proteins* **36**: 54–67, 1999.

Wall ME, Subramaniam S, Phillips Jr GN, Protein structure determination using a database of inter-atomic distance probabilities, *Protein Sci* **8**: 2720–2727, 1999.

Cui F, Jernigan R, Wu Z, Refinement of NMR-determined protein structures with database derived distance constraints, *J Bioinform Comput Biol* **3**: 1315–1329, 2005.

Cui F, Mukhopadhyay K, Young WB, Jernigan RL, Wu Z, Refinement of under-determined loops of human prion protein by database-derived distance constraints, *IEEE International Conference on Bioinformatics and Biomedicine Workshops*, pp. 14–23, 2007.

Wu D, Cui F, Jernigan R, Wu Z, PIDD: A database for protein inter-atomic distance distribution, *Nucleic Acids Res* **35**: D202–D207, 2007.

Wu D, Jernigan R, Wu Z, Refinement of NMR-determined protein structures with database derived mean force potentials, *Proteins Struct Funct Bioinform* **68**: 232–242, 2007.

Beyond structural computing

Chen L, Oughtred R, Berman HM, Westbrook J, TargetDB: A target registration database for structural genomics projects, *Bioinformatics* **20**: 2860–2862, 2004.

Xie L, Bourne P, Functional coverage of the human genome by existing structures, structural genomics targets and homology models, *PLoS Comput Biol* **1**: e31, 2005.

Kouranov A, Xie L, de la Cruz J, Chen L, Westbrook J, Bourne PE, Berman HM, The RCSB PDB information portal for structural genomics, *Nucleic Acids Res* **34**: D302–D305, 2006.

Chandonia J, Brenner SE, The impact of structural genomics: Expectations and outcomes, *Science* **311**: 347–351, 2006.

Kriete A, Eils R, *Computational Systems Biology*, Elsevier-Academic Press, 2005.

Palsson B, *Systems Biology — Properties of Reconstructed Networks*, Cambridge University Press, 2006.

Alon U, *An Introduction to Systems Biology: Design Principles of Biological Circuits*, CRC Press, 2006.

Werner E, The future and limits of systems biology, *Sci STKE* **5**: 16, 2005.

Appendix A

Design and Analysis of Computer Algorithms

In some sense, a computer algorithm is an abstraction of a computer program. An algorithm is a formal description of the computation to be conducted. It can be implemented into a program. A program is a set of instructions that can be executed on a computer. It can be written in different ways for the same algorithm. An algorithm provides only a high-level description of the work. A program contains detailed calculations and depends on the computer. However, a program can be described, analyzed, and evaluated easily at an algorithmic level without running on a specific computer.

An algorithm can be given at different levels. It can sometimes be as detailed as a program, but sometimes contain only a few mathematical formulas. In any case, to make a meaningful description on a computational task, an algorithm usually has input and output components, and can be carried out and terminated in a finite number of steps. An algorithm is often described in terms of some predefined primitive steps as well, so that the time and space required by the algorithm can be estimated based on its required number of primitive computational steps and memory units.

A.1. Evaluation of Algorithms

An algorithm can be evaluated in terms of the maximum computing time and memory space required for the solution of a problem. On the other hand, the difficulty of a problem can be assessed based on the minimum computing time and memory space required by an algorithm for solving it. The latter is often named as the time and space complexity of the problem.

A.1.1. *Computational model*

To be independent of a specific computer, an abstract computational model is often constructed as a common ground for analyzing or comparing algorithms. One of such models is the random access machine (RAM); another one is the Turing machine (TM). The RAM model is almost like a simplified computer. It is assumed to have a read-only input tape, a write-only output tape, and a memory. A program is a sequence of instructions that can be executed by the machine. The memory consists of a set of registers for holding temporary results. The final result can be written on the output tape. The instruction set typically constrains operations such as those in the following table:

Typical instruction set for RAM.

read x:	read a symbol from input tape to x.
write x:	write content in x on output tape.
load x:	move content in x to default register.
store x:	move content in default register to x.
add x:	add content in x to default register.
sub x:	subtract x from default register.
jump j:	jump to the jth instruction.
halt:	halt the machine.

Assume that on the input tape, the values of x, y, and z are provided in sequence. The calculation of $x - (y + z)$ can be carried out with a RAM program as follows:

Example RAM program.

read 1:	read x to register 1.
read 2:	read y to register 2.
read 3:	read z to register 3.
load 2:	copy y from register 2 to default register.
add 3:	add y and z.
store 2:	save $y + z$ to register 2.
load 1:	copy x from register 1 to default register.
sub 2:	subtract $y + z$ from x.
store 1:	save $x - (y + z)$ to register 1.
write 1:	write $x - (y + z)$.
halt:	halt the machine.

A typical Turing machine (TM) model has multiple read-and-write tapes. Its operation is determined by a finite control program. Formally, a TM is defined by a set of variables: $(Q, T, I, \delta, b, q_0, q_f)$, where Q is the set of states, T is the set of tape symbols, I is the set of input symbols (I is contained in T), b is the blank symbol, and q_0 and q_f are the initial and final states, respectively. The variable δ is the set of state transition functions mapping a subset of $Q \times T^*$ to $Q \times (T \times \{L, R, S\})^k$. For example, let $Q = \{q_0, q_1, q_2\}$, $T = \{0, 1, \#\}$, $I = \{0, 1\}$, $b = \{b\}$, and δ be defined such that

$$\delta(q_0, 0) = (q_1, (0, R)),$$
$$\delta(q_1, 1) = (q_1, (1, R)),$$
$$\delta(q_1, 0) = (q_2, (0, R)).$$

Then, a single-tape TM defined by $(Q, T, I, \delta, b, q_0, q_2)$ recognizes a sequence of $01\ldots10$.

In principle, the computing time for an algorithm can be evaluated by the number of basic RAM or TM operations required to run the algorithm on these machines. The memory space can be measured by the number of registers or the tape units the algorithm uses. However, in practice, higher-level units such as arithmetic operations for computing time and integers for memory space may be more convenient.

A.1.2. *Computing time*

The time that an algorithm takes to solve a problem can be used as a measure of the efficiency of the algorithm. However, the time is often machine-dependent and varies with different test problems. Therefore, it is more meaningful to compare the time on an abstract machine such as the ones we have discussed in the previous section and use a standard time unit such as the time to carry out some primary computational task. In practice, higher-level time units may also be used. For example, in numerical computing, the number of floating-point operations (including additions, subtractions, and multiplications) is often counted as the total computation time.

It also makes more sense to evaluate an algorithm on a class of problems instead of a specific one. The problems in the same class have the same nature, but often differ in their sizes. For example, systems of linear equations contain different numbers of equations, and the shortest path problems are formulated for different sizes of graphs. It is important to know the time for solving a problem of a given size in the worst case; it is also important to know how the time increases as the problem size increases. For this purpose, a number n is often used as an indicator for the size of the problem, such as the number of unknowns to be determined in a linear system of equations, and the time can be measured as a function of n.

Let $f(n)$ be the time that an algorithm solves a problem of size n. Then, $f(n)$ is said to be on the same order of $g(n)$ if there is a constant c such that $f(n) \leq c\,g(n)$ for sufficiently large n, and vice versa. The computation time of the algorithm is then said to be on the order of $g(n)$ or $f(n)$ and is denoted as $O(g(n))$ or $O(f(n))$, whichever is simpler. In this way, the algorithms requiring the same order of computation time for the solution of a given class of problems are considered to be equally efficient. On the other hand, if they require different orders of computation time, one may require a significantly longer time than another for the same size of problems, and the time may also increase faster as the problem size increases.

A.1.3. *Memory space*

An algorithm can also be evaluated based on the amount of memory space it requires for solving a given class of problems. To be independent of a specific machine, a standard amount of memory space such as that for storing an integer or a character can be used as the unit for space. If the problem size is n, the space can again be determined as a function of n. Let $f(n)$ be the space that an algorithm requires to solve a problem of size n. Then, $f(n)$ is said to be in the same order of $g(n)$ if there is a constant c such that $f(n) \leq c\, g(n)$ for sufficiently large n, and vice versa. The memory space of the algorithm is then said to be on the order of $g(n)$ or $f(n)$ and is denoted as $O(g(n))$ or $O(f(n))$, whichever is simpler. In general, the function for memory space is a lower bound of that for computation time, because the algorithm has to process the data in each memory unit at least once.

A.1.4. *Example analysis*

Example 1. Suppose that we want to find number a in a list of n ordered numbers. Let the space unit be defined as the space for storing one number. Then, the memory space for storing the whole list of numbers must be in order of n, i.e. $O(n)$. Let the ith number in the list be denoted as a_i. Then, the following algorithm can be used to search for a:

```
l = 0; u = n;
while l < u
    i = ceiling [(u − l) / 2];
    if a = aᵢ, stop;
    if a < aᵢ, u = i − 1; else l = i ;
end
```

Note that every time the algorithm enters the while loop, the list is cut in half. Let the length of the list be n and the total number of times the algorithm enters the loop be k. Then, $n / 2^k = 1$ and $k = \log_2 n$. Since in the while loop, the algorithm carries out only a few arithmetic operations, say c arithmetic operations, the total number of operations in the algorithm does not exceed $c \log_2 n$. Therefore, the computing time of the algorithm should be on the order of $\log_2 n$, i.e. $O(\log_2 n)$.

Example 2. Suppose that we want to merge two lists of ordered numbers, $\{a_i\}$ and $\{b_i\}$. Assume that the first list has n elements, and the second m. Then, at least $n + m$ memory units are required to store these numbers. Let $\{a_i\}$ be stored in n memory units and $\{b_i\}$ in m units. Let $\{c_i\}$ be the merged list and be saved in another set of $n + m$ memory units. The following algorithm then merges $\{a_i\}$ and $\{b_i\}$:

```
i = 1; j = 1; k = 1;
while i < n and j < m
    if a_i ≤ b_j
        c_k = a_i; i = i + 1;
    else
        c_k = b_j; j = j + 1;
    end
    k = k + 1;
end
while i < n
    c_k = a_i; i = i + 1; k = k + 1;
end
while j < m
    c_k = b_j; j = j + 1; k = k + 1;
end
```

Note that the total number of operations performed in the first loop does not exceed $n + m$ because each time the algorithm enters the loop, only one index is advanced, either i or j. Upon finishing the first loop, if the i index has not been advanced to n, the second loop is executed, and the total number of operations adds up to $n + m$; otherwise, if the j index has not been advanced to m, the third loop is executed instead, and the total number of operations also adds up to $n + m$. So, the computing time for this algorithm is $O(n + m)$. Assume that m is less than or equal to n. Then, the time can also be considered as on the order of $2n$, which is equivalent to $O(n)$.

A.2. Intractability

A problem can be easy or difficult to solve. Such a property can be characterized in terms of the computing time or memory space required to solve the problem. The latter is called the computational complexity of the problem or, more specifically, the time or space complexity of the problem. Based on their computational complexities, problems can be classified as tractable (easy) or intractable (hard) problems. Informally speaking, a tractable problem is a problem that can be solved in polynomial time or, in other words, in $O(f(n))$ computing time with $f(n)$ being a polynomial function of n. An intractable problem is a problem that any algorithm for the solution of the problem would take exponentially many steps in n in the worst case and cannot provide a solution to the problem in a reasonable time frame, especially when n is large.

A.2.1. *NP-completeness*

A problem is said to be NP-hard if it cannot be solved in polynomial time deterministically. An NP-hard problem is NP-complete if it can be solved in polynomial time on a nondeterministic machine. A nondeterministic machine is capable of always making right decisions and is therefore more powerful than a deterministic machine. However, current computers are all deterministic machines. If a sequence of decisions can be made in linear time on a nondeterministic machine, then on a regular computer (deterministic machine), exponentially many trials may be required to reach the right sequence of decisions.

Let P be the set of problems that can be solved in a deterministic polynomial time, and NP the set of problems that can be solved in a nondeterministic polynomial time. Then, P is a subset of NP. That is, a problem that can be solved in a deterministic polynomial time must be solvable in a nondeterministic polynomial time. However, it is not known whether the converse is true. It is assumed to be untrue. A common experience is that the solution of a problem in NP, but not in P, often requires an exponential time on a deterministic machine.

The problems that are in NP, but not in P, are called NP-complete problems.

A.2.2. *Satisfiability problem*

Let x be a Boolean variable representing a logical event. Then, $x = 1$ if the event occurs and $x = 0$ otherwise. Let $\bar{x}$ be the negation of x. Then, $\bar{x} = 1$ if $x = 0$ and $\bar{x} = 0$ if $x = 1$. Let xy be the logical "and" of x and y, i.e. $xy = 1$ if $x = 1$ and $y = 1$ and $xy = 0$ if $x = 0$ or $y = 0$; and $x + y$ be the logical "or" of x and y, i.e. $x + y = 1$ if $x = 1$ or $y = 1$ and $x + y = 0$ if $x = 0$ and $y = 0$. Then, a logic expression can be written as a mathematical formula using the above notations.

A logic expression can always be written in a standard conjunctive form, i.e. as a conjunction of clauses $C_1, C_2, \ldots, C_m$, where each clause is a disjunction of literals $l_1, l_2, \ldots, l_k$, with each literal being a variable or its negation. For example,

$$S = (\bar{x}_1 + x_2 + x_3)(x_1 + \bar{x}_2 + x_4)(x_2 + x_3 + \bar{x}_4),$$

where x_1, x_2, x_3, and x_4 are variables. An expression S is said to be satisfiable if there is an appropriate assignment of 0–1 values to the variables so that the expression is equal to 1. The problem of testing the satisfiability of a given logic expression is called the satisfiability problem. If the number of literals in each clause is restricted to three, the expression is said to be in a standard three-conjuctive normal form (3-CNF), and the corresponding satisfiability problem is called the three-satisfiability problem.

For a given logic expression of n variables, there can be 2^n possible assignments of 0–1 values to the variables. Therefore, the satisfiability problem cannot be solved straightforwardly by trying all possible assignments because, in the worst case, an exponentially many number of trials may need to be conducted, which will not be computationally feasible for large n. However, the problem is solvable in a nondeterministic polynomial time.

Theorem A.2.2.1. *The three-satisfiability problem or the satisfiability problem in general is NP-complete.*

A.2.3. *Set partition problem*

Let $S = \{a_1, a_2, \ldots, a_n\}$ be a given set of integers. A set partition problem is to partition the set S into two subsets S_1 and S_2 so that the sums of the integers in S_1 and S_2 are equal. Suppose S_1 has only one integer. Then, there can be n possible ways for such partitions, each with a different integer in S_1. In other words, there are C_n^1 (n choose 1) such partitions. Suppose S_1 has two integers. Then, there can be $n(n-1)/2$ possible ways for such partitions, each with a different pair of integers in S_1. In other words, there are C_n^2 (n choose 2) such partitions. With the same procedure, for partitions with $3, 4, \ldots, n$ integers in S_1, there can be $C_n^3, C_n^4, \ldots, C_n^n$ different ways for each correspondingly. Therefore, if the case for S_1 to be empty is also included, the total number of possible partitions for S would be

$$\sum_{i=0}^{n} C_n^i = 1 + \frac{n}{1!} + \frac{n(n-1)}{2!} + \cdots + \frac{n(n-1)(n-2)\cdots 1}{n!} = 2^n.$$

Clearly, the set partition problem cannot be solved straightforwardly by trying all possible partitions of the given set. In fact, it is impossible to find a deterministic polynomial time algorithm for the solution of the problem.

Theorem A.2.3.1.   *The integer set partition problem is NP-complete.*

A.2.4. *Polynomial time reduction*

A problem A is said to be polynomial-time reducible to another problem B if, in a polynomial time, problem A can be translated to problem B. If problem A can be reduced to problem B in polynomial time and if problem B can be solved in a polynomial time, then so can problem A; and if problem A is NP-hard, then problem B must be at least as hard as problem A.

Theorem A.2.4.1. *If problem A is polynomial-time reducible to problem B, problem A is polynomial-time solvable if problem B is, and problem B is NP-hard if problem A is.*

Based on the above theorem, whether or not a problem is NP-hard can often be verified by showing if a known NP-hard problem can be reduced to this problem. For example, a 0–1-integer linear programming problem is to find a 0–1-integer vector x such that $Ax \le b$ for given matrix A and vector b. We can show that the 0–1-integer linear programming problem is NP-hard because a known NP-hard problem such as the integer set partition problem or the three-satisfiability problem can be reduced to this problem, as demonstrated below.

Reduction from an integer set partition problem to a 0–1-integer linear programming problem: Let $S = \{a_1, a_2, \ldots, a_n\}$ be an integer set. Suppose that we want to partition S into two subsets S_1 and S_2 so that the sums of the integers in S_1 and S_2 are equal. Then, let $A = (a_1, a_2, \ldots, a_n)$ and $x = (x_1, x_2, \ldots, x_n)^\mathrm{T}$, and the set partition problem for S can be reduced to a problem for finding a 0–1-integer vector x such that $Ax = b/2$, where $b = \mathrm{sum}\{a_i : i = 1, 2, \ldots, n\}$, which is in fact a 0–1-integer linear programming problem. Therefore, the 0–1-integer linear programming problem cannot be solved in a deterministic polynomial time because the integer set partition problem would otherwise be polynomial-time solvable, contradicting the fact that it is actually NP-hard as stated in Theorem A.2.3.1.

Reduction from a three-satisfiability problem to a 0–1-integer linear programming problem: Let S be a logic expression, in a standard CNF form, $S = C_1 C_2 \cdots C_m$, where $C_i = li_1 + li_2 + li_3$, with li_j being any of the variables x_k or their negations y_k, $k = 1, 2, \ldots, n$. Suppose we want to see if we can find appropriate 0–1 values for variables x_k, $k = 1, 2, \ldots, n$ so that $S = 1$. Let $x = (x_1, x_2, \ldots, x_n)^\mathrm{T}$ and $y = (y_1, y_2, \ldots, y_n)^\mathrm{T}$. Then, $x_k + y_k = 1$, $k = 1, 2, \ldots, n$. Also, for each clause $C_i, li_1 + li_2 + li_3 \ge 1$ in order for $C_i = 1$, where $li_1, li_2,$ and li_3 are variables x_k or y_k, $k = 1, 2, \ldots, n$. Then, the three-satisfiability problem can be solved by finding 0–1 integer values for x and y so that the above constraints are all satisfied. The latter is in fact a 0–1-integer linear programming problem.

A.3. Lists, Arrays, Graphs, and Trees

A set of data can be stored either randomly or in a certain order or pattern. The data structure, i.e. the organization of the data, is important for the design of an efficient algorithm. Example data structures include lists, arrays, graphs, and trees.

A.3.1. *Lists*

A list is a set of data stored in a linear order. If the list contains n items, it can be stored in n memory units. Two pieces of information about the list, its beginning address and length, must be recorded so that the data in the list can be processed.

A new item can be added to a list: if the starting address of the list is s and the length of the list is n, then the new item is usually stored in unit $s + n$ and n is advanced by 1. The operation requires a constant time, i.e. $O(1)$.

An item in a list can be deleted: if the kth item is to be deleted, the item in unit $i+1$ is usually copied to unit i for $i = s+k-1, \ldots, s+n-1$, and n is reduced by 1. The operation requires an $O(n)$ computation time.

An item in a list can be searched: starting from the beginning unit, items in the unit are checked until the desired one is found. The operation requires an $O(n)$ computation time.

There are two special types of list with certain special operation rules: one is called a queue, and the other a stack. A queue is a list of data for which an item can be deleted only from the top of the list, while a new item can be added only at the bottom of the list. The deletion for a queue can be implemented by first removing the item in unit s and then copying the item in unit $i+1$ to unit i for $i = s+1, \ldots, s+n-1$ and finally reducing n to $n - 1$. The addition can be performed by simply storing the new item to unit $s+n$ and advancing n by 1. A stack is a list of data for which the data can only be deleted or added at the top of the list. If an addition is performed, the item in unit i is copied to unit $i+1$ for $i = s+n-1, \ldots, s$, and the new item is stored in unit s and n is advanced by 1. If a deletion is performed, the item in unit $i+1$ is copied to unit i for $i = s + 1, \ldots, s + n - 1$, and n is reduced by 1.

A.3.2. *Arrays*

Arrays are generalized lists. An array can have more than one dimension. The data in an array are specified by the indices along all of the dimensions. So, a one-dimensional (1D) array is a list, a two-dimensional (2D) array can be used to represent a matrix, a three-dimensional (3D) array corresponds to a 3D lattice, and so on and so forth. Physically, an array can still be implemented as a list with the elements in the array listed in a certain order. For example, the elements can be listed always along the first dimension with varying indices in other dimensions.

An array can be used to represent data with multiple indices. In particular, it is straightforward for storing numerical data such as matrices in arrays. For example, an $m \times n$ matrix can be stored in an $m \times n$ array with the (i, j) element in the matrix corresponding to the data in the (i, j) unit of the array. Let the array be denoted by A. The (i, j) element of the matrix for any i and j can be accessed directly in the array unit $A(i, j)$. Other types of data such as tables, texts, and images can also be represented directly using arrays.

A.3.3. *Graphs*

A graph can be used to represent a set of data and, in particular, the relationships among different pieces of data. For example, a dataset of 1000 cities contains the names of the cities $\{city_1, city_2, city_3, \ldots\}$ and the direct flight connections between certain pairs of cities $\{(city_1, city_2), (city_1, city_4), (city_2, city_5), \ldots\}$. Such a dataset can be represented straightforwardly using a graph.

A graph has a set of vertices V and edges E. Let $V = \{v_1, v_2, \ldots, v_n\}$ and $E = \{(v_i, v_j),$ for some v_i, v_j in $V\}$. A graph G with vertices in V and edges in E is denoted as $G = (V, E)$. If (v_i, v_j) in E are ordered pairs, i.e. $(v_i, v_j) \neq (v_j, v_i)$, the graph is called directed. If E contains all (v_i, v_j) pairs, the graph is called complete. For example, let $V = \{v_1, v_2, \ldots, v_6\}$ be a set of vertices representing six cities and $E = \{(v_1, v_2), (v_2, v_3), (v_3, v_4), (v_4, v_5), (v_5, v_6), (v_1, v_3), (v_4, v_6)\}$ be a set of edges between pairs of cities where there are direct flight connections. Then, a graph $G = (V, E)$ can be defined and plotted for the cities $v_1, v_2, \ldots, v_6$ and the flight information among them.

A graph $G = (V, E)$ can be represented by using an $n \times m$ matrix, where n is the number of vertices $|V|$ and m is the number of edges $|E|$. The matrix is defined such that the (i, j) element of the matrix is equal to 1 if the jth edge in E connects the ith vertex in V. This matrix is called the adjacent matrix of G and can be stored as a 2D array, as discussed in the previous section.

A basic operation on the data represented by a graph is to take a complete tour on the graph along the paths defined by the edges of the graph. The tour can be done in two ways: one is called the depth-first tour, and the other the breadth-first tour. In the depth-first tour, every time a vertex is visited, one of its connected and nonvisited vertices is selected for the next visit. If there are no more connected and nonvisited vertices, the tour backtracks to a previously visited vertex until all vertices are visited. In the breadth-first tour, all of the vertices connected to a previously visited vertex are visited first before any of the vertices connected to the current level of vertices is selected for the next visit.

A.3.4. *Trees*

A sequence of vertices, $v_i, v_k, v_l, v_j, \ldots$, is called a path if every pair of neighboring vertices in the sequence is connected by an edge in the graph. A sequence of vertices, $v_i, v_k, v_l, v_j, \ldots, v_m$, is called a circle if the sequence is a path and $v_i = v_m$. A graph is called connected if there is a path between every pair of vertices in the graph. A tree is a connected graph without a circle.

Define one of the vertices in the tree, say v_1, as the root. Then, for every vertex v_i in the tree, there is a unique path from the root to the vertex. Let the path be $v_1, \ldots, v_j, v_i$. Then, v_j is called the parent vertex of v_i, and v_i is called the child vertex of v_j. A vertex is called a leaf if it does not have child vertices.

A tree can be used to represent the hierarchical structure of data. For example, the family relationships within a large group of genetically related people (assuming no marriage within the group) can be represented by using a so-called family tree, with the oldest ancestor at the root, followed by the first generation of people, the second, and so on and so forth. Each vertex represents a person, with the vertices under it corresponding to all of his or her offspring, and the vertices along the path from the root to it corresponding to all of his or her direct ancestors.

Since the root has a unique path to each of the vertices in a tree, the search for a specific vertex can be guided by the path from the root to the vertex. For example, a person can be found quickly in the family tree if we know all of his or her direct ancestors, because the vertices corresponding to these ancestors form a path from the root to the vertex representing this person; and we can start with the root and travel down in the tree along this path to find the vertex for the person in the end of the path.

A.4. Sorting, Searching, and Optimization

Here, we describe a group of algorithms used in many computer applications. These include sorting the items in a list, searching for a specific item in a tree, solving the shortest path problem, and computing the minimal weight spanning tree for a graph.

A.4.1. *Sorting*

Let $\{a_1, a_2, \ldots, a_n\}$ be a set of items. Suppose the items can be ordered by a relation, i.e. $a_i < a_j$ or $a_j < a_i$ for $1 \leq i < j \leq m$. Suppose we want to rearrange the items so that they are in an increasing order, i.e. $a_i < a_j$ if $i < j$ for any pair of items a_i and a_j. A general sorting algorithm can be designed as follows. We first compare the first item in the given set of items with the rest of the items. If an item is found to be smaller than the first item, we switch their positions. In the end, the first item becomes the smallest one. We then compare the second item with the rest of the items. If an item is found to be smaller than the second item, we again switch their positions. In the end, the second item becomes the second smallest one. We keep doing such comparisons for the rest of the items until all of the items are positioned in an increasing order.

Let A be a list and a_i be stored in $A(i)$ initially. Then, the algorithm can be stated more formally as shown below. Note that there are two embedded loops in the algorithm; therefore, in the worst case, $O(n^2)$ operations (comparisons) are required.

Sorting algorithm 1

```
for i = 1, n − 1
    for j = i + 1, n
        if A(i) > A(j)
            T = A(j), A(j) = A(i), A(i) = T;
        end
    end
end
```

Another commonly used sorting algorithm is based on the idea of recursively sorting the subsets of items and then merging them together. First, every two neighboring items are merged into a set of two sorted items. Then, every two neighboring pairs of sorted items are merged into a set of four sorted items. The process continues and generates larger and larger subsets of sorted items until all of the items are merged into a single set of sorted items.

Let sort(A) be a function which takes a list of items A and returns a sorted one. Let merge(X,Y) be a function which takes two sorted sets of items X and Y and combines them into a single and sorted set of items, where X and Y are also represented as lists. Then, the sorting algorithm based on the idea described above can be given formally in a recursive form as shown below. Note that we first define the functions sort() and merge(), and then call sort(A) to recursively sort and merge the subsets of A. Without loss of generality, we assume that the length of the list $n = 2^k$ for some k.

```
function Y = sort(X)
n = length(X);
if n = 1, Y = X, return;
Y = merge(sort(X(1:n/2)), sort(X(n/2:n)));
```

```
function Z = merge(X,Y)
i = 1, j = 1, k = 1;
m = length(X); n = length(Y);
while i ≤ m and j ≤ n
    if X(i) < Y(j)
        Z(k) = X(i); k = k + 1; i = i + 1;
    else
        Z(k) = Y(j); k = k + 1; j = j + 1;
    end
end
if i < m
    for j = 0, m − i
        Z(k + j) = X(i + j);
    end
end
if j < n
    for i = 0, n − j
        Z(k + i) = Y(j + i);
    end
end
```

Sorting algorithm 2

$A = \text{sort}(A)$.

Note that if the length of X is m and the length of Y is n, the number of operations to merge X and Y will be $O(m + n)$. The recursive sorting algorithm can be viewed as having a number of levels: at the kth level, set A is divided into 2^k subsets, and they are grouped to pairs of subsets and then merged to 2^{k-1} subsets. At the kth level, merging all of the pairs of subsets requires $O(n)$ operations ($O(2n/2^k)$ for each merger, $O(n)$ for 2^{k-1} mergers), where n is the length of the entire set A. Since there can be only $\log_2 n$ levels, the total computing time for the algorithm is $O(n\log_2 n)$.

A.4.2. *Searching*

Suppose that we have a binary tree T with a root node v_0. Suppose that each node has a left child node denoted as lchild(v) and a right child node denoted as rchild(v). Suppose that we want to search the tree to find a node u. There can be two ways to conduct the search: (1) each time after node v is visited, node lchild(v) will be visited next (unless v is a leaf), and then the sibling of v is visited next; or (2) each time after node v is visited, all of the nodes in v's generation are visited first before their child nodes are visited. The first approach is called the depth-first search, and the second the breadth-first search.

The depth-first search can be implemented by storing the nodes to be visited in a stack. That is, each time a node is visited, it is removed from the stack, but its child nodes are added to the stack. Let S be a stack. Let first_item_from_stack (S, v) be a function to remove a node v from the top of the stack S, and add_item_to_stack (S,v) be a function to add a node v at the top of the stack S. Then, the depth-first search algorithm can be described as follows:

Depth-first search

```
add_item_to_stack (S, v₀);
while S ≠ Φ
    v = first_item_from_stack (S);
    if v = u
       stop;
    else
       add_item_to_stack (S, rchild(v));
       add_item_to_stack (S, lchild(v));
    end
end
```

In contrast, the breadth-first search can be implemented by storing the nodes to be visited in a queue. That is, each time a node is visited, it is removed from the queue, but its child nodes are added to the queue. Let Q be a queue. Let first_item_from_queue (Q, v) be a function to remove a node v from the beginning of Q, and add_item_to_queue

(Q, v) be a function to add a node at the end of Q. Then, the breadth-first search algorithm can be described as follows:

Breadth-first search

```
add_item_to_list (Q, v0);
while Q ≠ Φ
    v = first_item_from_queue (Q);
    if v = u
        stop;
    else
        add_item_to_queue (Q, lchild(v));
        add_item_to_queue (Q, rchild(v));
    end
end
```

A.4.3. *Solution to the shortest path problem*

Let $G = (V, E, W)$ be a graph, where V is the set of the nodes, E are the edges, and W are the weights on the edges. The shortest path problem is to find a path between two of the nodes in the graph, say v_1 and v_n, so that the sum of the weights along the path is the smallest among all possible paths between the two nodes.

Assume that G is a directed graph and that there is an edge (v_i, v_j) between v_i and v_j only if $i < j$. Then, the problem of finding the shortest path from v_1 to v_n can be solved by dynamically finding the shortest path first from v_1 to v_2, then to v_3, and eventually to v_n. Such an idea is called dynamic programming, and has been applied to many other applications including sequence alignment.

Let L_j be the length of the shortest path from node v_1 to node v_j. Let P_j be the node in the shortest path from v_1 to v_j directly preceding v_j. Then, the following algorithm gives a solution to the shortest path problem using dynamic programming:

Solution to the shortest path problem

```
for j = 2, ..., n
        Lj = w1,j; P0 = 1;
```

```
for i = 2, ..., j − 1
    if L_i + w_{i,j} < L_j
        L_j = L_i + w_{i,j}; P_j = i;
    end
end
```
end

Assume that, at the end of the algorithm, $P_n = v_l$, $P_l = v_k$, $P_k = v_j$, etc. Then, the shortest path from v_1 to v_n can be traced back from v_n to P_n, from v_l to P_l, from v_k to P_k, and so on and so forth. Note that each inner loop takes a constant time. Therefore, the total computing time is easily seen to be proportional to the total number of embedded loops, which is approximately equal to $n(n − 1)/2$, or $O(n^2)$.

A.4.4. *Minimal weight spanning tree*

Let $G = (V, E, W)$ be a weighted graph. A graph H is called a subgraph of G if $V(H)$ is in $V(G)$ and $E(H)$ is in $E(G)$. A subgraph H is said to be spanning if $V(H) = V(G)$. A spanning subgraph H is called a spanning tree if H is a tree. Here, we consider the problem of finding a minimum weight spanning tree for a given graph G.

There are two ways of solving this problem. First, we can start with an empty tree T; then, we can keep adding to the tree a new edge from G that has the least weight, as long as T remains a tree after the addition. Second, we can start with graph G and keep removing the most-weighted edge from the graph, as long as the graph remains connected. Based on these ideas, we can have the following two algorithms for finding the minimum weight spanning tree for a graph.

Minimum weight spanning tree 1

```
select v in V(G); V(T) = {v}; E(T) = {Ø};
while T is not spanning
    V(T) = V(T) + {v_j}, E(T) = E(T) + {(v_i, v_j)}, (v_i, v_j) = argmin {w_{i,j},
    v_i in V(T), v_j not in V(T)};
end
```

Depending on which data structure is used, the computing time of the algorithm can be different. In general, for each node, to find a minimum weight edge whose end node has not been extended may take $O(m)$ operations, where m is the number of edges of the given graph. Therefore, if there are n nodes, the total computing time will be $O(nm)$.

Minimum weight spanning tree 2

$V(T) = V(G)$; $E(T) = E(G)$;
while T has a circle C
    $E(T) = E(T) - \{(v_i, v_j)\}$, $(v_i, v_j) = \mathrm{argmax}\{w_{i,j}, (v_i, v_j) \text{ in } C\}$;
end

Finding a maximum weight edge in a circle is easy and takes $O(n)$ operations. However, it is not so trivial to find a circle for a graph. One way to find a circle in a graph is to make a complete tour on the graph, either using a depth-first or breadth-first order. During the tour, if a child node is generated or visited twice from a pair of different parent nodes, there must be a circle that can be found by tracing the two parent nodes back to their first common ancestor. Such a tour may take $O(m)$ steps in the worst case. Therefore, the entire algorithm may again take $O(mn)$ operations.

Selected Further Readings

Cormen TH, Leiserson CE, Rivest RL, Stein C, *Introduction to Algorithms*, The MIT Press, 2001.

Dasgupta S, Papadimitriou CH, Vazirani U, *Algorithms*, McGraw-Hill, 2006.

Aho AV, Ullman JD, Hopcroft JE, *Data Structures and Algorithms*, Addison Wesley, 1983.

Baldwin D, Hazzan O, *Algorithms and Data Structures: The Science of Computing*, Charles River Media, 2004.

Hopcroft JE, Motwani R, Ullman JD, *Introduction to Automata Theory, Languages, and Computation*, Addison Wesley, 2006.

Kozen DC, *Theory of Computation (Texts in Computer Science)*, Springer, 2006.

Cook WJ, Cunningham WH, Pulleyblank WR, Schrijver A, *Combinatorial Optimizations*, Wiley, 1997.

Wolsey LA, Nemhauser GL, *Integer and Combinatorial Optimization*, Wiley, 1999.

Appendix B

Numerical Methods

Numerical methods are the methods that can be used by computers for solving mathematical problems such as linear or nonlinear systems of equations, optimization problems, ordinary or partial differential equations, etc. Except for the linear cases, most of these problems can only be solved approximately. Therefore, numerical methods are often approximation methods in nature. In addition, on computers, all of the calculations are performed with finite precision. Therefore, some of the methods may even have trouble delivering expected solutions to given problems due to accumulated rounding errors.

B.1. Numerical Linear Algebra

Numerical linear algebra is a fundamental field of numerical computing, because the numerical solutions of many mathematical problems are eventually reduced to or supported by linear algebraic calculations. Numerical linear algebra includes the numerical methods for the solution of all linear algebraic problems.

B.1.1. *Matrix–vector operations*

The basic units in linear algebra are vectors and matrices. A vector is a list of numbers arranged in a row or column, called a row or column vector. A matrix is a set of numbers arranged in several rows and columns. A vector of n numbers is called an n-dimensional vector because it corresponds to a point in n-dimensional space, with its elements corresponding to the coordinates of the point in that space. A matrix of mn numbers arranged in m rows and n columns is called an $m \times n$ dimensional matrix, corresponding to a point in the product of m- and n-dimensional spaces. It can be considered to consist of m n-dimensional row vectors or n m-dimensional column vectors. An n-dimensional row vector can be considered as a $1 \times n$ dimensional matrix, while an n-dimensional column vector can be considered as an $n \times 1$ matrix. For convenience, vectors and matrices in our discussions are all assumed to be in real spaces unless specified explicitly otherwise.

There are several basic mathematical operations such as additions, subtractions, and multiplications that can be applied to vectors or matrices. First, let X be an $m \times n$ matrix. Then, X^{T} is called the transpose of X, and is a matrix obtained by changing the rows in the original matrix into columns. As a result, X^{T} is an $n \times m$ matrix. Since vectors are matrices, too, if x is a row vector, x^{T} becomes a column vector; and if x is a column vector; x^{T} becomes a row vector.

Now, let $y = (x_1, x_2, \ldots, x_n)^{\mathrm{T}}$ and $y = (y_1, y_2, \ldots, y_n)^{\mathrm{T}}$ be two column vectors of the same dimension. Then,

$$x \pm y = (x_1 \pm y_1, x_2 \pm y_2, \ldots, x_n \pm y_n)^{\mathrm{T}}. \tag{B.1}$$

If α is a scalar, then

$$\alpha x = (\alpha x_1, \alpha x_2, \ldots, \alpha x_n)^{\mathrm{T}}. \tag{B.2}$$

Let $\cdot$ represent the dot product of two vectors. Then,

$$x \cdot y = x_1 y_1 + x_2 y_2 + \cdots + x_n y_n. \tag{B.3}$$

The above operations can be defined for row vectors in the same fashion, but do not need to repeat. Let $X = \{x_{i,j} : i = 1, 2, \ldots, m, j = 1, 2, \ldots, n\}$ and $Y = \{y_{i,j} : i = 1, 2, \ldots, m, j = 1, 2, \ldots, n\}$ be two matrices with the same row and column dimensions. Then,

$$X \pm Y = \{x_{i,j} \pm y_{i,j} : i = 1, 2, \ldots, m, j = 1, 2, \ldots, n\}. \tag{B.4}$$

Again, if α is a scalar, then

$$\alpha X = \{\alpha x_{i,j} : i = 1, 2, \ldots, m, \; j = 1, 2, \ldots, n\}. \tag{B.5}$$

Now, let X be an $m \times l$ matrix and Y an $l \times n$ matrix. Then, the product of X and Y is an $m \times n$ matrix Z, and

$$z_{i,j} = x_{i,1} y_{1,j} + x_{i,2} y_{2,j} + \cdots + x_{i,l} y_{l,j}, \quad i = 1, 2, \ldots, m,$$
$$j = 1, 2, \ldots, n. \tag{B.6}$$

Note that when multiplying two matrices, the number of columns of the first matrix must be equal to the number of rows of the second matrix, and the resulting matrix has the same number of rows as the first matrix and the same number of columns as the second matrix. For this reason, not all matrices can multiply and the order of the matrices in the multiplication also matters.

Since vectors are also matrices, we can also multiply two vectors as long as they meet the dimensional requirement for matrix multiplication. Therefore, if x and y are two column vectors as defined above, we can have $x^\mathrm{T} y$, $y^\mathrm{T} x$, xy^T, and yx^T, where $x^\mathrm{T} y$ and $y^\mathrm{T} x$ are scalars and called the inner products of x and y,

$$x^\mathrm{T} y = y^\mathrm{T} x = x_1 y_1 + x_2 y_2 + \cdots + x_n y_n, \tag{B.7}$$

and xy^T and yx^T are matrices and called the outer products of x and y,

$$xy^\mathrm{T} = [yx^\mathrm{T}]^\mathrm{T} = \{x_i y_j : i = 1, 2, \ldots, n, \; j = 1, 2, \ldots, n\}. \tag{B.8}$$

However, we cannot have xy and yx, for if n is not equal to 1, the number of columns of the first matrix (which is equal to 1) is not the same as the number of rows of the second matrix (which is equal to n).

Several special types of matrices are worth mentioning. An $m \times n$ matrix is called a square matrix if m is equal to n. Two $n \times n$ square matrices can always multiply. However, the order is still important. For arbitrary two $n \times n$ matrices A and B, we have $AB \neq BA$ in general. A square matrix is called a lower triangular matrix if all of its upper off-diagonal elements are zero. Similarly, a square matrix is called an upper triangular matrix if all of its lower off-diagonal elements are zero. If all off-diagonal elements are zero, the matrix is called a diagonal matrix. A diagonal matrix is called an identity matrix and is denoted as I if all of the diagonal elements are equal to 1. The identity matrix has a special property that any matrix multiplied by the identity matrix of the same dimension, either from left or right, remains the same as the original matrix. A square matrix A is said to be nonsingular or invertible if there is a matrix B such that $AB = BA = I$. Matrix B is called the inverse of A and is denoted as A^{-1}.

Note that if we count the basic arithmetic operation such as addition, subtraction, or multiplication as one floating-point operation, the total number of floating-point operations for vector addition/subtraction, vector–scalar multiplication, and vector dot product are proportional to the length of the vector(s). In other words, they require $O(n)$ floating-point operations, where n is the length of the vector(s). The total number of floating-point operations for matrix addition/subtraction and matrix–scalar multiplication is proportional to the number of elements in the matrix (or matrices) and is $O(mn)$, where m and n are the row and column dimensions of the matrix (or matrices). However, multiplying an $m \times l$ matrix with an $l \times n$ matrix requires $O(mnl)$ floating-point operations. So, for $n \times n$ square matrices, the multiplication requires $O(n^3)$ floating-point operations.

B.1.2. *Matrix factorizations*

A matrix can be decomposed into two or more matrices, as an integer can be divided into two or more integer factors. Here, let us only consider the square matrices. Commonly seen matrix factorizations include LU factorization, Cholesky factorization, and SVD (singular value decomposition) factorization. Let A be an $n \times n$ matrix. Then, an LU factorization of A is such that $A = LU$, where L is a lower triangular matrix and U is an upper triangular matrix. If, in addition, A is symmetric and positive definite, A can be factorized as LL^{T}, where L is a lower triangular matrix; this factorization is called the Cholesky factorization. An SVD factorization of A is such that $A = U\Sigma V$, where U and V are orthogonal matrices and Σ is a diagonal matrix with positive diagonal elements. We discuss LU and Cholesky factorizations in greater detail in this section, and SVD factorization in Sec. B.1.4.

LU factorization requires an order of $2n^3/3$ floating-point operations for an $n \times n$ matrix, while Cholesky factorization requires an order of $n^3/3$. Therefore, the computation cost for the Cholesky factorization is only half of that for LU factorization. In any case, a given matrix may not always have an LU or Cholesky factorization. Theorem B.1.2.1 and Theorem B.1.2.2 describe the conditions under which the factorizations exist.

Theorem B.1.2.1. *If all of its n leading principal minors of an $n \times n$ real matrix A are nonsingular, then A has an LU factorization, $A = LU$, where L and U are the lower and upper triangular matrices, respectively.*

Theorem B.1.2.2. *If A is an $n \times n$ real, symmetric, and positive definite matrix, then A has a unique factorization, $A = LL^{\mathrm{T}}$, where L is a lower triangular matrix with positive diagonal elements.*

LU factorization

for $k = 1, 2, \ldots, n$
    $l_{k,k} u_{k,k} = a_{k,k} - \Sigma_{s<k} l_{k,s} u_{s,k}$;
    for $j = k + 1, 2, \ldots, n$
        $u_{k,j} = (a_{k,j} - \Sigma_{s<k} l_{k,s} u_{s,j})/l_{k,k}$;
    end
    for $i = k + 1, 2, \ldots, n$
        $l_{i,k} = (a_{i,k} - \Sigma_{s<k} l_{i,s} u_{s,k})/u_{k,k}$;
    end
end

Cholesky factorization

for $k = 1, 2, \ldots, n$
    $l_{k,k} = \mathrm{sqrt}(a_{k,k} - \Sigma_{s<k} l_{k,s} u_{s,k})$;
    for $i = k + 1, 2, \ldots, n$
        $l_{i,k} = (a_{i,k} - \Sigma_{s<k} l_{i,s} u_{s,k})/l_{k,k}$;
    end
end

Note that if A is an $n \times n$ matrix, $A = \{a_{ij}, i, j = 1, 2, \ldots, n\}$, a submatrix $A_k = \{a_{ij}, i, j = 1, 2, \ldots, k\}$ is called the kth principal minor of A. Note also that a matrix is called positive semidefinite if, for any vector $x \neq 0$, $x^{\mathrm{T}}Ax \geq 0$; and positive definite if, for any vector $x \neq 0$, $x^{\mathrm{T}}Ax > 0$.

B.1.3. *Linear systems of equations*

Consider a linear system of equations:

$$
\begin{aligned}
a_{11}x_1 + a_{12}x_2 + \cdots + a_{1n}x_n &= b_1 \\
a_{21}x_1 + a_{22}x_2 + \cdots + a_{2n}x_n &= b_2 \\
&\cdots \\
a_{n1}x_1 + a_{n2}x_2 + \cdots + a_{nn}x_n &= b_n,
\end{aligned}
\tag{B.9}
$$

where a_{ij} are coefficients, x_i are variables, and b_i are given constants, $i, j = 1, 2, \ldots, n$. Let $A = \{a_{ij} : i, j = 1, 2, \ldots, n\}$, $x = (x_1, x_2, \ldots, x_n)^{\mathrm{T}}$, and $b = (b_1, b_2, \ldots, n)^{\mathrm{T}}$. Then, the above system can be written in a more compact form:

$$
Ax = b.
\tag{B.10}
$$

Here, if matrix A is nonsingular, the equation can be solved by inverting matrix A and $x = A^{-1}b$. However, inverting a matrix is not numerically stable or, in other words, may not be accurate using numerical methods. Therefore, the above equation is often solved by factorizing the matrix first, and then solving two triangular systems. For example, let $A = LU$. Then,

$$
LUx = b.
\tag{B.11}
$$

Let $y = Ux$. We then have

$$
\begin{aligned}
Ly &= b, \\
Ux &= y.
\end{aligned}
\tag{B.12}
$$

By solving the above two systems, we can obtain the solution x for the original equation. The LU factorization requires an order of $2n^3/3$ floating-point operations. The solution of a triangular system can be obtained in an order of n^2 floating-point operations. Therefore, the computation cost for solving a linear system of equations is generally in an order of n^3 floating-point operations.

If matrix A is symmetric and positive definite, the equation can be solved by using the Cholesky factorization. Let the Cholesky factorization of A be $A = LL^{\mathrm{T}}$. Then,

$$LL^{\mathrm{T}}x = b. \tag{B.13}$$

Let $y = L^{\mathrm{T}}x$. We then have

$$\begin{aligned} Ly &= b, \\ L^{\mathrm{T}}x &= y. \end{aligned} \tag{B.14}$$

By solving the above two triangular systems, we obtain the solution x for the original equation.

Solution of lower triangular system

$r_i = b_i, i = 1, 2, \ldots, n;$
for $i = 1, 2, \ldots, n$
$\quad y_i = r_i / l_{ii};$
$\quad$ for $j = i + 1, i + 2, \ldots, n$
$\quad\quad r_j = r_j - l_{ji} y_i;$
$\quad$ end
end

Solution of upper triangular system

$r_i = y_i, i = 1, 2, \ldots, n;$
for $i = n, n - 1, \ldots, 1$
$\quad x_i = r_i / u_{ii};$
$\quad$ for $j = i - 1, i - 2, \ldots, 1$
$\quad\quad r_j = r_j - u_{ji} x_i;$
$\quad$ end
end

B.1.4. *Singular value decomposition*

Let A be an $m \times n$ matrix. Assume that $m \leq n$. Then, AA^{T} must be a symmetric and positive semidefinite matrix. Let $\sigma_1^2, \sigma_2^2, \ldots, \sigma_m^2$ be the eigenvalues of AA^T and $\sigma_1 \geq \sigma_2 \geq \cdots \geq \sigma_m$. Let u_i be the eigenvectors of AA^T corresponding to the eigenvalues σ_i^2, $i = 1, 2, \ldots, m$. Then,

$$AA^T u_i = \sigma_i^2 u_i, \quad i = 1, 2, \ldots, m. \tag{B.15}$$

Let $A^T u_i = \sigma_i v_i$. Then,

$$A v_i = \sigma_i u_i, \quad i = 1, 2, \ldots, m. \tag{B.16}$$

Define an $m \times m$ orthogonal matrix $U = [u_1, u_2, \ldots, u_m]$ and an $n \times n$ orthogonal matrix $V = [v_1, v_2, \ldots, v_m, \ldots, v_n]$ with additional orthogonal vectors $v_{m+1}, v_{m+2}, \ldots, v_n$. Let Σ be an $m \times n$ diagonal matrix with m diagonal elements $\sigma_1, \sigma_2, \ldots, \sigma_m$. Then,

$$AV = U\Sigma \quad or \quad A = U\Sigma V^T. \tag{B.17}$$

Here, $A = U\Sigma V^T$ is called a singular value decomposition of A. The diagonal elements $\sigma_1, \sigma_2, \ldots, \sigma_m$ are called the singular values of A. Several important properties related to the SVD of a matrix can be stated in the following theorems.

Theorem B.1.4.1. *An $m \times n$ matrix A can always be factorized as $A = U\Sigma V^T$, where U and V are $m \times m$ and $n \times n$ orthogonal matrices, respectively, and Σ is an $m \times n$ diagonal matrix with m nonnegative diagonal elements $\sigma_1, \sigma_2, \ldots, \sigma_m$.*

Theorem B.1.4.2. *Assume that the singular values of an $m \times n$ matrix A can be ordered in such a way that $\sigma_1 \geq \sigma_2 \geq \cdots \geq \sigma_r > \sigma_{r+1} = \cdots = \sigma_m = 0$. Then, the rank of A is equal to r, that is, the number of positive singular values of A.*

Theorem B.1.4.3. *Let A be an $m \times n$ matrix with singular value decomposition $A = U\Sigma V^T$. Define $A^+ = V\Sigma^{-1} U^T$. Then, A^+ is called the pseudoinverse of A, and $x = A^+ b$ minimizes $\|Ax - b\|$ or, in other words, solves a least squares problem for equation $Ax = b$.*

Theorem B.1.4.4. *Let A be an $m \times n$ matrix with singular value decomposition $A = U\Sigma V^{\mathrm{T}}$. Define Σ_k to be an $m \times n$ diagonal matrix with only first k nonzero diagonal elements of Σ. Then, $B = U\Sigma_k V^{\mathrm{T}}$ minimizes $\|A - B\|$ for all matrices B of rank k or, in other words, makes the best approximation to A by a matrix of rank k.*

B.2. Numerical Optimization

A general optimization problem can be stated as

$$\min_{x \in R^n} \ f(x)$$
$$\text{subject to } h_i(x) = 0, \quad i \in E; \quad g_i(x) \geq 0, \quad i \in I, \tag{B.18}$$

where $f : R^n \to R$ is called the objective function, $h_i : R^n \to R$ and $g_i : R^n \to R$ are called the constraint functions, and E and I are the index sets for the equality and inequality constraints, respectively.

Depending on the objective function and the constraints, an optimization problem can be linear or nonlinear, constrained or unconstrained, and discrete or continuous. If the objective and constraint functions are all linear, the problem is a linear optimization (or linear programming) problem. If there is no constraint, the problem is simply an unconstrained optimization problem. If the functions are defined in a continuous domain, the problem is a continuous optimization problem. It is a discrete optimization problem if some of the variables can only be selected from a discrete set of values such as the set of integers. The latter is called an integer optimization (or integer programming) problem. Here, we only discuss continuous, nonlinear, unconstrained optimization.

A general unconstrained optimization problem can be written as

$$\min_{x \in R^n} f(x), \tag{B.19}$$

where f is a nonlinear function. The problem may have two types of solutions called local or global solutions. A local solution to the problem, also called a local minimizer of f, is a point x^* in R^n such that $f(x^*) \leq f(x)$ for all x in an arbitrarily small neighborhood N of

x^*. A global solution to the problem, also called a global minimizer of f, is a point x^* in R^n such that $f(x^*) \leq f(x)$ for all x in R^n. A local optimal solution is relatively easy to obtain. A global optimal solution is generally difficult to find.

B.2.1. *Steepest descent direction method*

A necessary condition for a point x^* in R^n to be a local optimal solution to an unconstrained optimization problem is that the gradient of the objective function at x^* is equal to zero. The condition is also sufficient if, in addition, the Hessian of the objective function at x^* is positive definite.

Theorem B.2.1.1. *A point x^* in R^n is a local minimizer of f only if the gradient of f at x^* is equal to zero, i.e. $\nabla f(x^*) = 0$.*

Theorem B.2.1.2. *A point x^* in R^n is a local minimizer of f if the gradient of f at x^* is equal to zero and the Hessian is positive definite, i.e. $y^T \nabla^2 f(x^*) y > 0$ for all $y \neq 0$.*

The steepest descent direction method for local optimization is a method for finding a local minimizer of a given objective function f. It generates a sequence of points $\{x_k\}$ that converges to a point x^* so that $\nabla f(x^*) = 0$. At every step, it generates a new point x_{k+1} from a previous point x_k along a direction p_k. The direction p_k is a decreasing direction of f at x_k if $\nabla f(x_k)^T p_k < 0$. The function f decreases the most when p_k is in $-\nabla f(x_k)$ direction. Therefore, the steepest descent direction method always chooses the most decreasing direction to generate the next point. That is, at every step,

$$x_{k+1} = x_k + \alpha_k p_k, \tag{B.20}$$

where $p_k = -\nabla f(x_k)/\|\nabla f(x_k)\|$ and α_k is a properly selected step length. The process of selecting the step length α_k, called line search, is critical for the sequence of iterates to eventually converge to a local minimizer of the objective function. Exact line search is to find a step length so that the function can be decreased the most along p_k. That is, in exact line search, α_k is selected to be the solution to the following one-dimensional (1D) optimization problem:

$$\min_{\alpha_k} f(x_k + \alpha_k p_k). \tag{B.21}$$

Note that, in general, the direction p_k does not have to be the most descent and the step length α_k does not have to be the optimal length. In fact, as long as p_k is a descent direction and α_k makes the function decrease sufficiently, the sequence of iterates will converge. We call such a method the descent direction method.

Let p_k be a descent direction and α_k a step length. We say that α_k satisfies the Wolfe conditions if

$$f(x_k + \alpha_k p_k) \leq f(x_k) + c_1 \alpha_k \nabla f(x_k)^{\mathrm{T}} p_k, \tag{B.22}$$

$$\nabla f(x_k + \alpha_k p_k)^{\mathrm{T}} p_k \geq c_2 \nabla f(x_k)^{\mathrm{T}} p_k, \tag{B.23}$$

where $0 < c_1 < 1$ and $c_1 < c_2 < 1$. The following theorems show that if the objective function is bounded below and is continuously differentiable, a step length that satisfies the Wolfe conditions can always be found along a descent direction; and if such a step length is used in the descent direction method, the sequence of iterates generated by the method can be guaranteed to converge to a local minimizer of the objective function.

Theorem B.2.1.3. *Suppose that f is bounded below and is continuously differentiable. Suppose that p_k is a descent direction. Then, there always exists an interval (α_l, α_u) with $0 < \alpha_l < \alpha_u < \infty$ so that the Wolfe conditions hold for all α_k in (α_l, α_u).*

Theorem B.2.1.4. *Suppose that f is bounded below and is continuously differentiable. Let p_k be a descent direction and α_k be selected by using line search and satisfy the Wolfe conditions. Then, the sequence of iterates generated by the descent direction method converges to a local minimizer of f.*

B.2.2. *Conjugate gradient method*

Consider a quadratic optimization problem,

$$\min_{x \in R^n} x^\mathrm{T} A x / 2 - b^\mathrm{T} x, \qquad \text{(B.24)}$$

and assume that A is symmetric and positive definite. Based on Theorem B.2.1.2, a point x in R^n is a solution to the quadratic optimization problem (B.24) if and only if it satisfies the condition

$$A x = b. \qquad \text{(B.25)}$$

Therefore, the solution to the quadratic optimization problem (B.24) can be obtained by directly solving Eq. (B.25). Alternatively, the solution to problem (B.24) can also be obtained by applying an optimization method, say the steepest descent direction method, to the objective function of the problem. The former is called the direct method, and the latter the iterative method.

A conjugate direction method is an iterative method for the solution of a quadratic optimization problem as given in (B.24). A set of directions $\{p_k : k = 1, 2, \ldots, n\}$ is said to be conjugate with respect to a matrix A if

$$p_i^\mathrm{T} A p_j = \begin{cases} 0, & i \neq j \\ 1, & i = j. \end{cases} \qquad \text{(B.26)}$$

It is not hard to verify that the directions $\{p_k\}$ are linearly independent, given that A is symmetric and positive definite. A conjugate direction method for the solution of the quadratic optimization problem (B.24) is to generate a sequence of iterates $\{x_k\}$ by using a set of descent and conjugate directions $\{p_k\}$. Such a sequence of iterates can converge to a solution to the problem in n steps.

Theorem B.2.2.1. *Let $\{p_k : k = 1, 2, \ldots, n\}$ be a set of descent directions and be conjugate with respect to A. Let $\{x_k\}$ be a sequence of iterates generated by the formula $x_{k+1} = x_k + \alpha_k p_k$, where α_k is determined by using exact line search. Then, $\{x_k\}$ converges to a solution to the quadratic optimization problem (B.24) in, at most, n steps.*

A set of conjugate directions may not necessarily be a set of descent directions. A method called the conjugate gradient method is therefore developed to generate a set of directions that are guaranteed to be both descent and conjugate. In this method, first, an initial descent direction is selected; then, at each of the following steps, a new direction p_{k+1} is determined using a linear combination of the previous direction p_k and the negative gradient direction $-g_k$. Let b_{k+1} be a constant. Then, $p_{k+1} = p_k - b_{k+1}g_k$, where b_{k+1} is determined so that p_{k+1} and p_k are conjugate with respect to A.

Linear conjugate gradient method

$x_0 = 0; r_0 = b - Ax_0; p_0 = r_0;$
for $k = 0, 1, \ldots, n$
$\qquad x_{k+1} = x_k + a_k p_k;$
$\qquad r_{k+1} = b - Ax_k;$
$\qquad b_{k+1} = r_{k+1}^{\mathrm{T}} r_{k+1} / r_k^{\mathrm{T}} r_k;$
$\qquad p_{k+1} = p_k + b_{k+1} r_k;$
end

Theorem B.2.2.2. *The sequences of iterates and directions generated by the conjugate gradient method have the following properties: (1) $\{p_k\}$ are descent directions. (2) $\{p_k\}$ are conjugate with respect to A. (3) $Ax_n = b$.*

Theorem B.2.2.2 simply shows that the conjugate gradient method generates a sequence of descent and conjugate directions; and by Theorem B.2.2.1, the sequence of iterates generated by using these directions converges to the solution of the quadratic optimization problem (B.24) in n steps. The direct method for the solution of the quadratic optimization problem (B.24) requires $O(n^3)$ floating-point operations. The conjugate gradient method requires the same amount of computing time in the worst-case scenario when n steps are taken. However, the method may sometimes take fewer than n steps to finish and, therefore, require less computing time than the direct method.

Because of its advantage, the conjugate gradient method has also been extended to the solution of general nonlinear optimization problems. If the general objective function is close to a strictly convex quadratic function, the method may be able to find the solution for the problem as efficiently as it does for a strictly convex quadratic function. In fact, nearby a local minimizer, the objective function is indeed close to a strictly convex quadratic function, and fast convergence of the conjugate gradient method to the local minimizer can in general be expected.

Nonlinear conjugate gradient method

$x_0 = 0; g_0 = \nabla f(x_0); p_0 = -g_0;$
for $k = 1, 2, \ldots$
    $x_{k+1} = x_k + a_k p_k;$
    $g_{k+1} = \nabla f(x_{k+1});$
    $b_{k+1} = g_{k+1}^T g_{k+1} / g_k^T g_k;$
    $p_{k+1} = p_k - b_{k+1} g_{k+1};$

end

B.2.3. *Newton method*

Let f be the objective function and ∇f be the gradient. Then, in the Newton method, in order to minimize f, the equation $\nabla f(x) = 0$ is solved to obtain a critical point of f. Similar to other approaches, a sequence of points $\{x_k\}$ is generated to converge to the critical point or, in other words, the solution of the equation $\nabla f(x) = 0$. The sequence $\{x_k\}$ is generated as follows. At every step, at x_k, the gradient ∇f is approximated by a linear function,

$$\nabla f(x) \approx \nabla f(x_k) + \nabla^2 f(x_k)(x - x_k). \tag{B.27}$$

Then, x_{k+1} is obtained from the solution of the equation

$$\nabla f(x_k) + \nabla^2 f(x_k)(x - x_k) = 0, \tag{B.28}$$

that is,

$$x_{k+1} = x_k - [\nabla^2 f(x_k)]^{-1} \nabla f(x_k). \tag{B.29}$$

Let $g_k = \nabla f(x_k)$ and $H_k = \nabla^2 f(x_k)$. Then,

$$x_{k+1} = x_k - H_k^{-1} g_k, \tag{B.30}$$

where $s_k = -H_k^{-1} g_k$ is called the Newton direction or Newton step and can be obtained by solving a linear system of equations,

$$H_k s_k = -g_k. \tag{B.31}$$

The Newton direction may not necessarily be a descent direction. Therefore, the sequence of iterates generated by the Newton method may not converge. However, if the starting point is close to a local minimizer of f, the sequence will converge in general, and the convergence rate is greater than that by the steepest descent method or the conjugate gradient method.

Theorem B.2.3.1. *Let f be twice continuously differentiable. Let x^* be a local minimizer of f, and assume that the Hessian of f at x^* is positive definite. Then, the sequence of iterates generated by the Newton method converges to x^* locally quadratically, provided the starting point x_0 is sufficiently close to x^*.*

Note that the steepest descent method and the conjugate gradient method converge only linearly and, in general, are slower than the Newton method. However, the Newton method requires the solution of a linear system of equations at every step and is computationally more expensive. Here, by converge linearly, we mean that

$$||x_{k+1} - x^*|| \leq K||x_k - x^*||, \tag{B.32}$$

for $0 < K < 1$ and k sufficiently large; and by converge quadratically, we mean that

$$||x_{k+1} - x^*|| \leq K||x_k - x^*||^2, \tag{B.33}$$

for $K > 0$ and k sufficiently large. We also say that the convergence is superlinear if

$$||x_{k+1} - x^*|| \leq K||x_k - x^*||, \tag{B.34}$$

for k sufficiently large and $\lim_{k \to \infty} K = 0$.

B.2.4. *Quasi-Newton method*

The quasi-Newton method is a generalization of the secant method for the solution of a single equation to the solution of a system of equations. So, at every step, an approximated Hessian B_k is used to obtain the linear approximation to the gradient, and

$$x_{k+1} = x_k - \alpha_k B_k^{-1} g_k, \tag{B.35}$$

where α_k is the step length. The approximated Hessian B_{k+1} is usually obtained by making a low rank update to B_k so that a general secant equation

$$B_{k+1} s_k = y_k \tag{B.36}$$

is satisfied, where $s_k = x_{k+1} - x_k$ and $y_k = g_{k+1} - g_k$.

Several update formulas have been used to obtain the Hessian approximation or its inverse. The first set of formulas is called the symmetric rank-1 update formulas (SR1). Let B_k and H_k be the Hessian approximation and its inverse at step k. Then,

$$(SR1) \quad B_{k+1} = B_k + \frac{(y_k - B_k s_k)(y_k - B_k s_k)}{(y_k - B_k s_k)^{\mathrm{T}} s_k}, \tag{B.37}$$

$$(SR1) \quad H_{k+1} = H_k + \frac{(s_k - H_k y_k)(s_k - H_k y_k)}{(s_k - H_k y_k)^{\mathrm{T}} y_k}. \tag{B.38}$$

It is easy to verify that if B_k and H_k are symmetric, B_{k+1} and H_{k+1} are also symmetric, and they satisfy the following secant equations:

$$B_{k+1} s_k = y_k,$$
$$\tag{B.39}$$
$$H_{k+1} y_k = s_k.$$

Theorem B.2.4.1. *The matrices B_{k+1} and H_{k+1} obtained by using the symmetric rank-1 update formulas preserve the nonsingularity of the matrices B_k and H_k if and only if $(y_k - B_k s_k)^{\mathrm{T}} s_k \neq 0$ and $(s_k - H_k y_k)^{\mathrm{T}} y_k \neq 0$.*

Another set of update formulas are the Davidon-Fletcher-Powell (DFP) formulas:

$$(DFP) \quad B_{k+1} = (I - \rho_k y_k s_k^{\mathrm{T}}) B_k (I - \rho_k s_k y_k^{\mathrm{T}}) + \rho_k y_k y_k^{\mathrm{T}}, \tag{B.40}$$

$$(DFP) \quad H_{k+1} = H_k - \frac{H_k y_k y_k^{\mathrm{T}} H_k}{y_k^{\mathrm{T}} H_k y_k} + \frac{s_k s_k^{\mathrm{T}}}{y_k^{\mathrm{T}} s_k}, \tag{B.41}$$

where $\rho_k = 1/y_k^T s_k$. A similar set of formulas are the Broyden-Fletcher-Goldfarb-Shanno (BFGS) formulas:

$$(BFGS) \quad B_{k+1} = B_k - \frac{B_k s_k s_k^T B_k}{s_k^T B_k s_k} + \frac{y_k y_k^T}{y_k^T s_k}, \tag{B.42}$$

$$(BFGS) \quad H_{k+1} = (I - \rho_k s_k y_k^T) H_k (I - \rho_k y_k s_k^T) + \rho_k s_k s_k^T. \tag{B.43}$$

Both DFP and BFGS formulas preserve not only the symmetry, but also the positive definiteness of the matrices as long as $\rho_k > 0$. Therefore, they are more commonly used in practice. Several important convergence results for the quasi-Newton methods can be obtained, as stated in the following theorems.

Theorem B.2.4.2. *Let f be a strictly convex quadratic function, with a positive definite Hessian. Let $B_0 = I$, and B_k be obtained by using the BFGS formula, for all $k > 0$. Then, the sequence of iterates $\{x_k\}$ obtained by the quasi-Newton method with exact line search converges to a local minimizer x^* of f in n steps.*

Theorem B.2.4.3. *Let f be a twice continuously differentiable function. Let $B_0 = I$, and B_k be obtained by using the BFGS formula, for all $k > 0$. Suppose that line search is used and the Wolfe conditions are satisfied by all of the step lengths. Then, the sequence of iterates $\{x_k\}$ obtained by the quasi-Newton method converges to a local minimizer x^* of f superlinearly.*

B.3. Numerical Solutions to Initial-Value Problems

We consider the following initial-value problem and describe the general numerical methods for its solution:

$$x' = f(t, x), \quad x(0) = x_0, \tag{B.44}$$

where x is a function of t and f is a function of t and x. The equation contains the first derivative of x, and is therefore called a first-order equation. The theory and methods for this equation can often be extended to higher-order systems of equations, which we will not cover.

B.3.1. *Existence and uniqueness of the solution*

In general, a first-order differential equation has infinitely many solutions, but it will have a single solution once an initial condition is specified. The following theorems provide a theoretical foundation for this property.

Theorem B.3.1.1. *If f and $\partial f/\partial x$ are continuous and f has a maximum M for all t and x such that $|t - t_0| \leq a$ and $|x - x_0| \leq b$, then the initial-value problem (B.44) has a unique solution $x(t)$ defined for all t such that $|t - t_0| \leq \min(a, b/M)$.*

Theorem B.3.1.2. *If f is continuous for all t and x such that $a \leq t - t_0 \leq b$ and $c \leq x \leq d$, and if $|f(t, x_1) - f(t, x_2)| \leq L|x_1 - x_2|$ for any x_1 and x_2 and constant L, then the initial-value problem (B.44) has a unique solution $x(t)$ defined for all t such that $a \leq t - t_0 \leq b$.*

B.3.2. *Single-step method*

The simplest numerical method for solving an initial-value problem is the Euler method. The method computes a sequence of points on the solution trajectory, beginning with the position given by the initial condition. Let $x(t)$ be the function value at time t. The method computes the function value $x(t + h)$ at time $t + h$ based on a linear approximation of the function at t,

$$x(t + h) = x(t) + x'(t)h. \tag{B.45}$$

Since $x'(t) = f(x, t)$, the formula can also be written as

$$x(t + h) = x(t) + f(t, x)h. \tag{B.46}$$

Depending on the problem, the step size h cannot be too large; otherwise, the approximation will not be accurate enough.

Since the Euler method is based on linear approximation, the error due to truncation is in the order of h^2. In order to increase the accuracy, it is natural to use a higher-order approximation, say the second- or third-order approximation. For example, the second derivative of x can be obtained by differentiating the equation

$$x''(t) = f_x(t, x)x'(t) + f_t(t, x) = f_x(t, x)f(t, x) + f_t(t, x), \qquad \text{(B.47)}$$

and $x(t + h)$ can therefore be calculated using the formula

$$x(t + h) = x(t) + f(t, x)h + \frac{[f_x(t, x)f(t, x) + f_t(t, x)]h^2}{2}, \qquad \text{(B.48)}$$

with a higher order of accuracy, $O(h^3)$. Here, the calculations of f_x and f_t are of course required, but by making another approximation

$$f(t + h, x + x'h) = f(t, x) + f_x(t, x)x'h + f_t(t, x)h, \qquad \text{(B.49)}$$

the formula can be changed to depend on f only:

$$x(t + h) = x(t) + \frac{[f(t, x) + f(t + h, x + f(t, x)h)]h}{2}. \qquad \text{(B.50)}$$

This formula is called the second-order Runge–Kutta method or Heun's method.

B.3.3. *Multistep method*

The methods described in the previous section are single-step methods because they only use the knowledge on $x(t)$ for the calculation of $x(t + h)$. In other words, if t_k is the time and x_k is the function value at step k, the calculation of x_{k+1} at step $k + 1$ depends only on x_k. Instead, a multistep method tries to compute x_{k+1} using information on x_k and $x_{k-1}, x_{k-2}, \ldots$ at a number of previous steps. The idea is to use the following equation to compute x_{k+1}:

$$x_{k+1} = x_k + \int_{t_k}^{t_{k+1}} f(t, x(t))dt, \qquad \text{(B.51)}$$

where the integral can be calculated numerically based on the values of f at a set of previous steps $k, k - 1, k - 2, \ldots$, etc. Suppose that the resulting formula has the form

$$x_{k+1} = x_k + af_k + bf_{k-1} + cf_{k-2} + \cdots, \qquad \text{(B.52)}$$

where f_k is the function value for f at x_k and t_k, and $a, b, c, \ldots$ are constants determined by numerical integration rules. Such a formula is called the Adams–Bashforth formula. The following is an Adams–Bashforth formula of order 5:

$$x_{k+1} = x_k + \frac{1}{720}[1901 f_k - 2774 f_{k-1} + 2616 f_{k-2}$$
$$- 1274 f_{k-3} + 251 f_{k-4}]. \tag{B.53}$$

If the function value f_{k+1} is also used, the formula for calculating x_{k+1} can be like

$$x_{k+1} = x_k + af_{k+1} + bf_k + cf_{k-1} + \cdots. \tag{B.54}$$

Such a formula is called the Adams–Moulton formula and can be applied along with an Adams–Bashforth formula. That is, the Adam–Bashforth formula is first used to compute a value for x_{k+1}, and the Adams–Moulton formula is then used to obtain f_{k+1} and a new value for x_{k+1}. Such a procedure is called a prediction-correction procedure and is considered to be more stable and accurate. The Adams–Bashforth is called the predictor, and the Adams–Moulton the corrector. The following is an Adams–Moulton formula of order 5:

$$x_{k+1} = x_k + \frac{1}{720}[251 f_{k+1} + 646 f_k - 264 f_{k-1}$$
$$+ 106 f_{k-2} - 19 f_{k-3}]. \tag{B.55}$$

B.3.4. *Accuracy and convergence*

Consider a numerical method for the solution of an initial-value problem given by the formula

$$a_m x_k + a_{m-1} x_{k-1} + \cdots + a_0 x_{k-m}$$
$$= h[b_m f_k + b_{m-1} f_{k-1} + \cdots + b_0 f_{k-m}], \tag{B.56}$$

where a_i and b_i are constants. The method is called explicit if $b_m = 0$, and implicit if $b_m \neq 0$. Associated with the formula are two polynomials:

$$p(z) = a_m z^m + a_{m-1} z^{m-1} + \cdots + a_0,$$
$$q(z) = b_m z^m + b_{m-1} z^{m-1} + \cdots + b_0. \tag{B.57}$$

Let $x(h, t)$ be the solution obtained by using Eq. (B.56) with a step size h. The method is said to be convergent if $\lim_{h\to 0} x(h, t) = x(t)$ for all t. The method is said to be stable if all roots of p lie in the disk $|z| \leq 1$ and if each root of modulus one is simple. The method is said to be consistent if $p(1) = 0$ and $p'(1) = q(1)$.

Theorem B.3.4.1. *For a numerical method in Eq. (B.56) for the solution of an initial-value problem to be convergent, it is necessary and sufficient that it is stable and consistent.*

Let $x(t)$ be the solution of an initial value problem. Let x_k be the value of x at t_k obtained using a numerical method in Eq. (B.56). Then, $x(t_k) - x_k$ is defined as the local truncation error of the method. For example, the Euler method has a truncation error $O(h^2)$. The accumulation of the local errors in the calculation of a sequence of points on the solution trajectory results in the global truncation error.

Theorem B.3.4.2. *In general, if the local truncation error of a numerical method in Eq. (B.56) for the solution of an initial-value problem is $O(h^{m+1})$, the global truncation error must be $O(h^m)$.*

B.4. Numerical Solutions to Boundary-Value Problems

The problem of finding the solution to a differential equation with a given set of conditions at the beginning and ending states of the solution is called a boundary-value problem. We consider the boundary-value problem for a second-order differential equation in the following form:

$$x'' = f(t, x, x'), \quad x(t_0) = x_0, \quad x(t_e) = x_e, \tag{B.58}$$

where there are two separate conditions at the beginning and ending states of the solution. The problem with such a set of conditions is also called a two-point boundary-value problem.

B.4.1. *Existence and uniqueness of the solution*

The theory and methods for initial-value problems cannot be directly applied to boundary-value problems. The conditions for a boundary-value problem to have a solution are more restrictive than for an initial-value problem. Sometimes, a boundary-value problem may not even have a solution, while the solution to an initial-value problem always exists under some modest conditions. In general, we have the following theorem.

Theorem B.4.1.1. *The two-point boundary-value problem*

$$x'' = f(t, x, x'), \quad x(t_0) = x_0, \quad x(t_e) = x_e \tag{B.59}$$

has a unique solution on interval $[t_0, t_e]$ provided that the function f and its partial derivatives f_t, f_x, $f_{x'}$ are all continuous on $D = [t_0, t_e] \times R \times R$, f_x is positive, and $f_{x'}$ is bounded on D.

A simpler class of problems than that in (B.59) is that

$$x'' = f(t, x), \quad x(t_0) = x_0, \quad x(t_e) = x_e. \tag{B.60}$$

It can be verified that, with a change of variable, this class of problem can be written in a special form:

$$x'' = f(t, x), \quad x(0) = 0, \quad x(1) = 0. \tag{B.61}$$

Theorem B.4.1.2. *The two-point boundary-value problem*

$$x'' = f(t, x), \quad x(0) = 0, \quad x(1) = 0 \tag{B.62}$$

has a unique solution on interval $[t_0, t_e]$ if f_x is continuous, nonnegative, and bounded on $D = [t_0, t_e] \times R$.

B.4.2. *Single shooting*

A simple approach to the boundary-value problem is to solve an initial-value problem for the equation with some beginning conditions so that the solution satisfies the given ending conditions. This is similar to shooting a ball in a basket (ending condition) from a given position (beginning condition). By adjusting the initial velocity (another beginning condition), one hopes to make a shot so that the ball can end in the basket (satisfying the ending condition).

Consider the boundary-value problem

$$x'' = f(t, x, x'), \quad x(t_0) = x_0, \quad x(t_e) = x_e, \tag{B.63}$$

and the corresponding initial-value problem

$$x'' = f(t, x, x'), \quad x(t_0) = x_0, \quad x'(t_0) = z. \tag{B.64}$$

We wish to find an appropriate z so that the corresponding solution $x(t; z)$ for the initial-value problem in (B.64) satisfies the conditions $x(t_0; z) = x_0$ and $x(t_e; z) = x_e$ specified in problem (B.63). Let $\varphi(z) = x(t_e; z) - x_e$. Then, we can find such a value for z by solving a nonlinear equation $\varphi(z) = 0$.

The equation $\varphi(z) = 0$ can be solved by using, for example, the Newton method; and at every step k,

$$z_{k+1} = z_k - \frac{\varphi(z_k)}{\varphi'(z_k)}. \tag{B.65}$$

Here, the calculation of $\varphi(z)$ requires the solution $x(t; z)$ of the initial-value problem in (B.64) evaluated at t_e, namely, $x(t_e; z)$. The derivative $\varphi'(z)$ is equal to $\partial x / \partial z$ evaluated at t_e, which can be computed either through finite-difference approximation or by solving another associated initial-value problem called the first variation equation. The equation can be obtained by differentiating Eq. (B.64) with respect to z:

$$\frac{\partial x''}{\partial z} = f_x \frac{\partial x}{\partial z} + f_{x'} \frac{\partial x'}{\partial z}, \quad \frac{\partial x}{\partial z}(t_0) = 0, \quad \frac{\partial x'}{\partial z}(t_0) = 1. \tag{B.66}$$

Let $y = \partial x / \partial z$. The above equation can be written as

$$y'' = f_x y + f_{x'} y', \quad y(t_0) = 0, \quad y'(t_0) = 1, \tag{B.67}$$

and be solved in conjunction with the solution to the initial-value problem in (B.64).

B.4.3. *Multiple shooting*

The single-shooting method is not stable, as we can imagine that if the time period is long, the "shot" will be sensitive to the change in the initial derivative and be difficult to reach the target. To overcome this difficulty, we can divide the time interval into smaller subintervals and make the shootings separately in the subintervals. Here, the function values and derivatives of the solution trajectory on the intermediate points are, of course, all unknown. We need to determine them so that the trajectories obtained in the subintervals can be connected into a smooth trajectory on the entire interval. The method for obtaining such a trajectory is called the multiple-shooting method.

We first divide the time interval $[t_0, t_e]$ uniformly into n subintervals $[t_0, t_1], [t_1, t_2], \ldots, [t_{n-1}, t_n]$, with $t_n = t_e$. Then, on each subinterval $[t_i, t_{i+1}]$, $0 \leq i < n$, we solve an initial-value problem,

$$x'' = f(t, x, x'), \quad x(t_i) = r_i, \quad x'(t_i) = s_i, \quad t_i \leq t \leq t_{i+1}. \tag{B.68}$$

Let the solution be denoted by $x_i(t; r_i; s_i)$. Then, in the entire interval,

$$x(t) \equiv x_i(t; r_i; s_i), \quad t_i \leq t \leq t_{i+1}, \quad 0 \leq i < n. \tag{B.69}$$

Here, we need to find $r = [r_0; r_1; \ldots; r_{n-1}]$ and $s = [s_0; s_1; \ldots; s_{n-1}]$ such that

$$x_i(t_{i+1}; r_i; s_i) = r_{i+1}, \quad x_i'(t_{i+1}; r_i; s_i) = s_{i+1}, \quad 0 \leq i < n - 2,$$
$$\tag{B.70}$$
$$x_0(t_0; r_0; s_0) = x_0, \quad x_{n-1}(t_n; r_{n-1}; s_{n-1}) = x_e.$$

Let $r = (r_0, r_1, r_2, \ldots, r_n)^{\mathrm{T}}$ and $s = (s_0, s_1, s_2, \ldots, s_n)^{\mathrm{T}}$. If we define a vector function μ such that

$$\mu(r; s) \equiv \begin{bmatrix} x_0(t_1; r_0; s_0) - r_1 \\ x_0'(t_1; r_0; s_0) - s_1 \\ \vdots \\ x_{n-2}(t_{n-1}; r_{n-2}; s_{n-2}) - r_{n-1} \\ x_{n-2}'(t_{n-1}; r_{n-2}; s_{n-2}) - s_{n-1} \\ x_0(t_0; r_0; s_0) - x_0 \\ x_{n-1}(t_n; r_{n-1}; s_{n-1}) - x_e \end{bmatrix}, \tag{B.71}$$

then we basically need to determine $(r; s)$ such that $\mu(r; s) = 0$. The equation can be solved again by using, for example, the Newton method, as we described in the previous section.

B.4.4. *Finite difference*

The finite-difference approach to a boundary-value problem is also based on the division of the time interval into a set of subintervals. Let $[t_0, t_e]$ be uniformly divided into $n + 1$ subintervals $[t_0, t_1], [t_1, t_2], \ldots, [t_n, t_{n+1}]$, with $t_{n+1} = t_e$. Then, on each subinterval $[t_i, t_{i+1}]$, $0 \leq i \leq n$, a difference equation can be obtained by making a difference approximation to each of the derivatives of the solution:

$$\frac{x_{i+1} - 2x_i + x_{i-1}}{h^2} = f\left(t_i, x_i, \frac{x_{i+1} - x_{i-1}}{2h}\right), \quad i = 1, 2, \ldots, n. \quad \text{(B.72)}$$

The values x_i, $i = 1, 2, \ldots, n$, can then be obtained by solving the difference equations simultaneously along with the given values x_0 and $x_{n+1} = x_e$ at t_0 and $t_{n+1} = t_e$. Note that this approach does not require the solution of initial-value problems as required in the shooting methods; all it needs is the solution of a system of algebraic equations.

Let $x = (x_1, x_2, \ldots, x_n)^{\mathrm{T}}$. Let $A = \{a_{ij}, i, j = 1, 2, \ldots, n\}$ with $a_{ij} = 2$ if $|i - j| = 0$, $a_{ij} = -1$ if $|i - j| = 1$, and $a_{ij} = 0$ otherwise, and

$$A = \begin{bmatrix} 2 & -1 & 0 & \cdots & 0 \\ -1 & 2 & -1 & 0 & \vdots \\ 0 & \ddots & \ddots & \ddots & 0 \\ \vdots & 0 & -1 & 2 & -1 \\ 0 & \cdots & 0 & -1 & 2 \end{bmatrix}. \quad \text{(B.73)}$$

Define a vector function F such that

$$F_i(x) = f\left(t_i, x_i, \frac{x_{i+1} - x_{i-1}}{2h}\right), \quad i = 2, 3, \ldots, n - 1,$$

$$F_1(x) = -\frac{x_0}{h^2} + f\left(t_1, x_1, \frac{x_2 - x_0}{2h}\right), \quad \text{(B.74)}$$

$$F_n(x) = -\frac{x_{n+1}}{h^2} + f\left(t_n, x_n, \frac{x_{n+1} - x_{n-1}}{2h}\right).$$

Then, Eq. (B.72) can be put into a compact form:

$$\frac{1}{h^2}Ax + F(x) = 0. \tag{B.75}$$

Stated in the following theorems are some important properties of the finite-difference method.

Theorem B.4.4.1. *Let f be continuously differentiable, f_x and $f_{x'}$ be bounded, and $f_{x'}$ be positive on $D = [t_0, t_e] \times R \times R$. Then, the system of difference equation (B.75) has a unique solution for any $h \leq 2/\max\{|f_x|\}$.*

Theorem B.4.4.2. *Let f be continuously differentiable, f_x and $f_{x'}$ be bounded, and $f_{x'}$ be positive on $D = [t_0, t_e] \times R \times R$. Then, the numerical solution $\{x_j\}$ obtained by the finite-difference method has an order of $O(h^2)$ local truncation error.*

Selected Further Readings

Kincaid DR, Cheney EW, *Numerical Analysis: Mathematics of Scientific Computing*, Brooks Cole, 2001.

Stoer J, Bulirsch R, Bartels R, Gautschi W, *Introduction to Numerical Analysis*, Springer, 2004.

Ortega JM, Rheinboldt WC, *Iterative Solution of Nonlinear Equations in Several Variables*, SIAM, 1987.

Dennis JE, Schnabel RB, *Numerical Methods for Unconstrained Optimization and Nonlinear Equations*, SIAM, 1996.

Fletcher R, *Practical Methods of Optimization*, Wiley, 2000.

Nocedal J, Wright S, *Numerical Optimization*, Springer, 2006.

Ascher U, Mattheij R, Russell R, *Numerical Solution of Boundary Value Problems for Ordinary Differential Equations*, SIAM, 1995.

Deuflhard P, Bornemann F, *Scientific Computing with Ordinary Differential Equations*, Springer, 2002.

Index